Praise for *A Shape in*

"If anyone has earned the right to write a book about brow
Alaska born, raised in a land shaped and defined by their presence, he's the perfect
guide on a journey that weaves natural and human history together with his own, often
harrowing experiences in bear country. Unflinchingly authentic, intensely personal,
artfully told, and deeply moving, *A Shape in the Dark* ranks as not only one of the best
bear books, but one of the greatest Alaska books ever."

—Nick Jans, author of *The Grizzly Maze* and *A Wolf Called Romeo*

"*A Shape in the Dark* is an at times terrifying, but always thoughtful and vivid portrayal
of America's storied history with brown bears. You'll find yourself listening for a twig
to snap or alder to sway as Bjorn Dihle transports you to another world, one where
you are not at the top of the food chain. These tales will stick with you long after you
close the book."

—Alex Messenger, author of *The Twenty-Ninth Day:*
Surviving a Grizzly Attack in the Canadian Tundra

"Wilderness guide Dihle creates a wide-ranging portrait of brown bears in this
adventurous collection of essays. The book, Dihle writes, is 'about our relationship
with brown bears,' though it's also an ode to Alaskan wilderness as his home state
becomes 'more encroached upon.' His experiences run between panic and reverence
in the face of the 'incredibly muscled and poised' animal, and interspersed with his
encounters are profiles of past adventurers and their relationships with bears. . . . He
creates memorable portraits of fellow explorers, but where Dihle's writing shines is
in his unwavering appreciation of and commitment to preserving bears' wild habitat:
'Once it's gone, it's gone forever.' With its vivid prose, this moving homage to Alaska
and those who live there really hits home."

—*Publishers Weekly*

"*A Shape in the Dark* . . . reveals the complicated morality around brown bears, wilderness,
and ideas of masculinity. It is reverent in discussing wilderness while acknowledging
human beings' enduring compulsions to subdue it. Rather than suggesting answers,
the book acknowledges human paradoxes and is humble about living within them."

—*Foreword Reviews*

"To walk in the land of the mighty grizzly bear is to delve into the wild of our being. Bjorn
Dihle takes us on this journey, exploring those who walked before him, and praying that
those who follow can discover their own true nature."

—Amy Gulick, author of *The Salmon Way: An Alaska State of Mind*

Bjorn Dihle

A SHAPE in the DARK

LIVING AND DYING WITH BROWN BEARS

MOUNTAINEERS
BOOKS

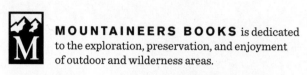

MOUNTAINEERS BOOKS is dedicated to the exploration, preservation, and enjoyment of outdoor and wilderness areas.

1001 SW Klickitat Way, Suite 201, Seattle, WA 98134
800-553-4453, www.mountaineersbooks.org

Printed in Canada
Distributed in the United Kingdom by Cordee, www.cordee.co.uk
24 23 22 21   1 2 3 4 5

Copyeditor: Ali Shaw
Cover and book design: Jen Grable
Cartographer: Martha Bostwick
Back cover photo: Bjorn Dihle

Portions of some chapters were previously published in a different form in *Alaska Magazine*, *Sierra Magazine*, and the *Juneau Empire*.

Library of Congress Cataloging-in-Publication Data is available at https://lccn.loc .gov/2020027843. The ebook record is available at https://lccn.loc.gov/2020027844.

Mountaineers Books titles may be purchased for corporate, educational, or other promotional sales, and our authors are available for a wide range of events. For information on special discounts or booking an author, contact our customer service at 800-553-4453 or mbooks@mountaineersbooks.org.

♻ Printed on 100% recycled and FSC-certified materials

ISBN (paperback): 978-1-68051-309-7
ISBN (ebook): 978-1-68051-310-3

An independent nonprofit publisher since 1960

MIX
Paper from
responsible sources
FSC
www.fsc.org
FSC® C016245

For MC and Shiras

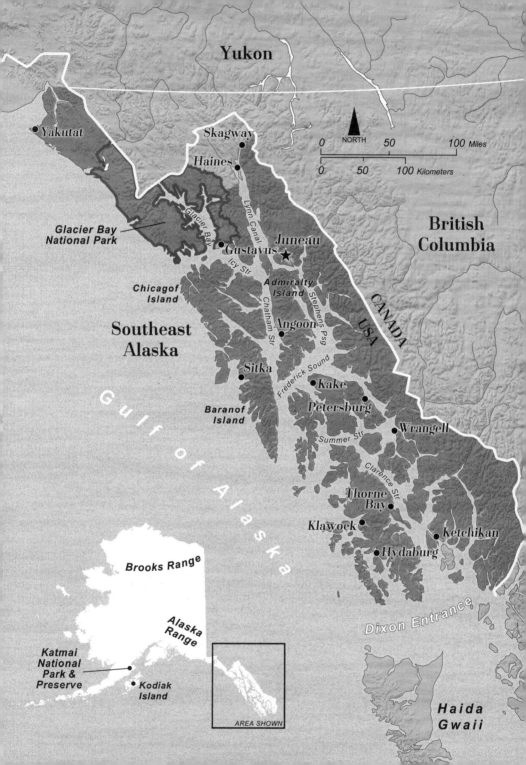

CONTENTS

PROLOGUE

In August of 2004, I borrowed my friend Forest Wagner's ancient pickup truck in Fairbanks and headed north. I had a couple weeks before my classes would begin at the University of Alaska Fairbanks, and I hoped to climb Mount Doonerak in the central Brooks Range. As I drove through graveyards of charred, spindly black spruce rising in the smoke from forest fires, the Alaska pipeline appeared every so often in the gray. The silhouette of the Brooks Range shone hazy in the smoke. I hid Forest's truck in a willow thicket ten miles south of the Continental Divide, shouldered my backpack, and began hiking up an ember-tinted valley. The muddy creek banks were lined with grizzly tracks and caribou bones. I traveled into the mountains, past herds of Dall sheep that watched me warily. At my second camp, I woke to a gray wolf howling from a nearby knoll. Two dark wolves, the wild amber of their eyes electrifying the soft glow of the early morning, walked by within yards of my tent.

I wandered deeper into the country, following caribou trails past antler sheds lying on the tundra—a mute testament to life in the heavy silence. Two wolves halfheartedly stalked a large Dall sheep ram as I watched in the distance. When I tried to climb Doonerak, I was thwarted by a series of steep, crumbly ridges and cliffs. On my second attempt, I followed a ridge that became too exposed and technical. September came. I was going to be late for classes, but still I made one last attempt on Doonerak. I found a route that, though it crossed some dangerously steep scree slopes, got me to the base of the summit. The top was an hour away and appeared to be a relatively easy rock scramble, but black clouds blew in and snow began to fall.

Hours later, as the storm broke and temperatures skyrocketed back close to seventy degrees, I sat outside my tent wishing I hadn't turned back. I didn't want to return to Fairbanks, but I was out of food. A porcupine waddled by, and I leapt to my feet and grabbed a large rock. I crept up behind the animal, raised the rock over my head, and then, hesitated. Rugged mountains rose indifferently into the pale blue sky as I watched the porcupine continue until it disappeared into the tundra. I broke camp and began the trek out.

Late that night the temperature plummeted, the wind picked up, and snow began to fall. By morning, a full storm battered my tent. I unzipped the door a crack and peered out to see a herd of several dozen sheep bedded nearby in the lee of the blizzard. In the early afternoon I broke camp and trudged through a four-inch blanket of snow. In a low valley, I watched a grizzly grazing on the last of the year's blueberries. It was smaller than the coastal brown bears I was familiar with, brutally muscled, and unaware of my presence. I edged past into thick willows with wind hard on my face. Once I was several hundred yards away, since I was traveling through brush with a lot of bear sign, I began singing 1980s power ballads to warn other animals. An hour later, while taking a few minutes' break from making noise, I emerged into a small clearing and nearly stepped on a grizzly.

For a second, before I remembered to be afraid, I was overcome by its sheer presence. Dark brown with silver forelegs, incredibly muscled and poised, it appeared too real to be real. A moment later, our eyes met. Fear and rage flashed in its small brown eyes. It closed the distance separating us in one stride as I reached for my bear spray attached to my pack strap. There wasn't time to unclip it. There wasn't even time to realize the bear was about to knock me down and what that might mean. But instead of smashing me, the bear stopped short—at what felt like only inches away—and recoiled to the side. I took a step back and was fumbling with my pepper spray when the bear came again and, just before contact seemed inevitable, bounded away. For the longest seconds of my life, we engaged in a strange and violent dance until the bear crashed off into the willows. It was only then that I had the pepper spray in my hand and ready. During the rest of the hike out, every set of grizzly tracks I came across seemed to radiate with the promise of death. At any moment, I expected a bear to emerge from the mountains and come for me.

At Coldfoot, the one gas station between Fairbanks and Prudhoe Bay, I picked up two Japanese hitchhikers who'd come to Fairbanks in search of the northern lights. In most of the Interior that summer, visibility was limited to a couple hundred yards due to smoke from forest fires. Instead of leaving defeated, they'd hitchhiked north a few hundred miles, far enough from the fires to where they made out a faint green glimmer in the sky one night.

"We see aurora borealis. Then we wait on the side of the road for three days! No one would pick us up. So cold! Now we know why they call it Coldfoot!" the younger of the two said. He was close to my age. Much to his parents' disapproval, he'd decided to travel for a year before finishing a college degree in economics.

"I don't want to go back to Japan," he said. "My family wants me to go to school, get married, and work, work, work."

The other man was older and possessed a calm intensity. His English wasn't good, but I learned that he'd been traveling for more than a decade.

"I never go back Japan," he said. "I just go."

We rode through thick smoke and smoldering black spruce without seeing another vehicle for over an hour. At one point small fires burned along both sides of the road, and I began to worry I might get us killed. When I mentioned that I was thinking about turning back to Coldfoot, they encouraged me to keep driving south. I dropped them off in downtown Fairbanks at a hostel.

At the university I fumbled through the proper motions, stared at the eastern Alaska mountain range, and felt lost. Instead of renting a cabin or room, I pitched my tent in the woods a few miles from campus. October came. With the first snow, the fires and smoke vanished, revealing blue sky, stars, and the occasional display of northern lights. At night I listened to small forest animals, or silence, or the panting and grunting of a rutting bull moose, or the soft hissing of falling snow, or the screeching of a lynx, and I thought of the bear.

I thought of how close it had come to maiming, even killing, me. I thought of how the threat of death made me realize how deeply in love with life I was. I woke at the smallest changes in the forest and wondered if the bear had traveled three hundred miles south to find me. In the morning I watched my "pet" mouse raiding the food bag and wondered what the bear was doing at that exact moment. What was it thinking? What was it feeling? Did it remember me? Was it in its winter den yet? I tried to apply myself to my studies, but there seemed little of truth or worth to be found. I missed the wild expanse where words, concepts, and beliefs meant nothing. I missed the wind and open horizon. I missed the mountains and the tundra. Most of all, I missed the bear.

INTRODUCTION

On October 1, 2018, my two brothers and I packed the meat, bones, and hide of a mountain goat down a mountain in northern Southeast Alaska. Near the ocean, we walked through an estuary that had been rototilled by hungry bears digging roots. The salmon runs and berries had failed that year. We paused to get water from a stream as the sunset illuminated mountains and ocean in soft layers of red. A fresh set of brown bear tracks crossed in the sand. Its humanlike hind track was a little over a foot long, and its pigeon-toed front track was over eight inches wide. I guessed it would stand more than eight feet on its hind legs and weigh close to eight hundred pounds—a large male in this region. Using our headlamps, we traveled through darkness, following the same game trail the bear was traveling through the forest. I thought how by now, up on the mountain, all that remained of the goat was a stain of blood and offal, and a few bones that had already been picked clean by eagles and ravens. Soon, a wolf, bear, or wolverine would crunch them to feast on marrow, and then the broken remains would be reabsorbed by the mountain. My light illuminated gnarled trees rising into the black, and I yelled out a warning to the bear. I knew it could hear me, and sensing its restlessness, I thanked it for allowing my brothers and me to pass through the woods.

Late that night, after I made it home, I learned that a young man had been killed by a brown bear on nearby Admiralty Island that morning. The man had been an eighteen-year-old Oklahoman who had just arrived in Alaska and begun a job as a driller's assistant. He and another worker had been helicoptered to a remote drilling site, leased by the Hecla Greens Creek Mining Company, high in the mountains on the island. The press release and newspaper articles that followed had few details as to what happened. I would hear later from people with ties to the mine that the young man had walked away from the drill pad to check a hose or look for water. Neither of the men had been armed. When the Oklahoman didn't return, the other contractor went looking and found a female bear with two big cubs eating him. The man radioed for help, summoning a helicopter, which tried to buzz the bears off the body. The bears wouldn't leave and were shot several times with a .375 rifle from the air, killing them.

This book is about our relationship with brown bears, which, in case there's any confusion, is the same species as the grizzly bear. During the last twenty years, I've spent a significant amount of time exploring wild places across Alaska and, for most of the last decade, worked as a guide taking people and film crews throughout Southeast Alaska to watch and film brown bears. Before and even after beginning this project, there have been times I almost hated bears. Like most feelings of hostility, mine were rooted in fear. Yet, there is no place I love more than grizzly country, and no animal has intrigued and challenged me more than the bear.

I grew up in Juneau and live on nearby Douglas Island, both of which are just a few miles' boat ride from Admiralty Island. The roughly hundred-mile-by-seventeen-mile island lies in the northern section of the Alexander Archipelago, a group of over a thousand rainforest islands in Southeast Alaska. The archipelago and the adjacent rugged mainland comprise the Tongass National Forest. Established by President Theodore Roosevelt in 1907, the Tongass is wet, wild, and America's largest national forest at nearly 26,500 square miles. Admiralty is part of the ABC Islands (Admiralty, Baranof, Chichagof, and—though most folks have never heard of them—Yakobi and Kruzof Islands), which make up the northern part of the Alexander Archipelago. These five islands are ecologically similar— historically, they have big runs of salmon and a lot of brown bears and Sitka black-tailed deer. The true name of Admiralty is Kootznoowoo, which in Lingít, the language of the Tlingit people who've lived in Southeast Alaska since time immemorial, translates to something like "fortress of the bears."

Admiralty is known for its exceptionally dark, even black, brown bears. These black grizzlies were once commonly called Shiras bears—and used as a rallying cry for conservationists who fought to keep the island from being clear-cut. For the most part, the only people who still use the term *Shiras* are old-timers who remember the battle for Admiralty. The island's bear population is roughly estimated to be one per square mile, meaning that it likely has more brown bears than the entire Lower 48.

It had been nearly a century since someone had been killed on Admiralty by a bear. Both deaths could have been avoided and were the result of people having little understanding of bears. Nearly every documented brown bear fatality in Southeast Alaska has occurred late in the year when bears enter a state of hyperphagia. (The one exception is a garbage bear that ate a man in Hyder, a tiny community near Alaska's southern border with Canada, in July of 2000.) Each fall, when the salmon runs peter out, an internal switch goes off in bears that makes them more voracious and, often, agitated. On Admiralty and neighboring islands, they spend more time in the high country feeding on berries and anything else with caloric value. A bear has to gain around 30 percent of what it weighed in the spring or it won't survive the long winter sleep.

Far fewer people die at the teeth and claws of brown bears in Alaska than most people realize. There were no documented fatalities in the state from 2006 until the fall of 2012, when two men died in separate attacks. There were no fatalities from 2013 until the summer of 2018, when a man was killed in Eagle River, near Anchorage. In contrast, nearly 2,000 of Alaska's estimated population of 30,000 to 40,000 brown bears are killed by sport hunters annually. In the Kodiak Archipelago alone, 600 miles west of Admiralty Island, $5 million is spent on bear hunts each year, and around 180 bears are killed. There's been only one documented case of a bear killing someone there in the last 75 years, so the demon monster mythology often applied to bears and perpetuated by media is far from the truth. Brown bears are generally not aggressive toward people unless threatened or surprised at close quarters, and those encounters are usually relatively easy to avoid. Still, there's no way to make bears safe, although people have tried and failed.

I PUT MY SHARE OF GOAT MEAT AND BONES IN THE REFRIGERATOR, TOOK a beer into the shower, and then lay in bed unable to sleep. My body was exhausted, but my mind wouldn't turn off. The residue of the goat's blood smelled subtle and sweet. My partner, MC, pregnant with our first child, snored gently next to me. Our golden retriever, Fen, shifted and laid her head on my foot. Still unable to sleep, I moved to the living room and stared out at the black of a mountain rising from the ocean in the darkness. The lights of Juneau shone weakly to the north across Gastineau Channel. I imagined the big bear whose tracks my brothers and I had followed earlier resting at the base of an ancient tree or prowling the night in search of something to ease its hunger. I thought of the fetus that was my kid, floating in amniotic fluid and protected from the world by a few thin layers of MC's flesh. I thought of the Oklahoman's parents and felt a tightening in my guts.

Pouring whiskey into a mason jar, I remembered how the fall before, while deer hunting on Admiralty Island, I had come upon a den on a steep slope beneath a giant spruce tree. A bear had recently excavated piles of dirt. Chances were, the animal lay feet away, asleep atop a bed of moss in the darkness. The hair on the back of my neck stood up. I quietly chambered a round in my rifle and held it to my shoulder. Taking a step back, I glanced about the shadowy rainforest. I imagined the bear rushing forth from the earth and charging down the mountain away from me. Then I imagined it rushing forth from the earth, knocking me down, and eating me. Nothing but my breathing and heartbeat disturbed the quiet. Then I crept closer and peered inside the den, like a moth drawn toward a flame.

I studied the darkness inside my condo and then looked out at the darkness outside the window. After a while, much like our ancestors who built fires to keep away the monsters, I opened my laptop and stared at the lit-up screen, hoping the words would come.

Part I

CHAPTER 1

The Skin of the Bear

This animal is the monarch of the country which he inhabits. The African lion or the tiger of Bengal are not more terrible or fierce. He is the enemy of man and literally thirsts for human blood. So far from shunning, he seldom fails to attack and even to hunt him. The Indians make war upon these ferocious monsters with the same ceremonies as they do upon a tribe of their own species, and, in the recital of their victories, the death of one of them gives the warrior greater renown than the scalp of a human enemy.

—Henry Marie Brackenridge, *Views of Louisiana*

I first encountered a brown bear when I was four or five years old. It was lying in a salmon stream on Admiralty Island, reduced to bones, pieces of hide, and tendrils of flesh. Thick salmonberry brush and giant trees rose above each bank, offering the illusion of impenetrable walls. A bald eagle glided past and landed on a logjam. Ravens spoke their ancient language from the boughs of Sitka spruce trees. Pink salmon filled the stream and struggled to spawn. Leaving the bear, I waded downstream toward an old cabin my family was staying at. I paused to throw rocks at a belly-up salmon who'd just laid the last of her eggs as she drifted downstream, toward the ocean.

I found my dad and took him back to the carcass, where he cradled his rifle and studied the surrounding salmonberry bushes and jungle. Dad had moved to Alaska from California when he was in his early twenties so he could hunt and experience wilderness. The brown bear has always been a big, mysterious animal to him, something that you have to respect and watch out for. The carcass that lay at our feet wasn't much bigger than me though. It was likely a two-year-old, the typical age for a cub to be run off by its mother so she could mate again. In the day or two since it had been killed, other bears, eagles, and ravens had reduced it to little more than a skeleton and a tattering of flesh. The stream's current and the falling rain added to the forces, encouraging blood from flesh and flesh from bone. In the shallows, on the gravel bars, and on the banks nearby were the carcasses of countless decomposing salmon. Even the living salmon, splashing against the current, jockeying for position and digging out redds, were in the process of dying.

I stood beside my dad with little understanding of what was going on. I can't recall my thoughts, but maybe I suffered a glimmer of awareness. Maybe I imagined a bear crashing through the woods, engulfing me in its jaws, and how even if I escaped its grip, eventually—like the spawning salmon—my blood, flesh, and guts would still be freed of my bones and carried away from the thing I thought was me. We left the carcass, and I followed my dad downstream, staring up at the brush and alders lining the banks, where bear trails led into the unknown.

IN OCTOBER OF 1809, MERIWETHER LEWIS TRAVELED BY HORSEBACK through the Tennessee hills, bound for Washington, DC, to report to President James Madison. Eight years prior, just after taking office, President Thomas Jefferson had handpicked Lewis to be his personal secretary. Jefferson dreamed of building a country that stretched from the Atlantic to the Pacific, and he believed that Lewis was the man to set the groundwork to make it a reality. The two became extremely close, their relationship taking on the dynamic of a wise and visionary father grooming his prodigy son for a quest of mythic proportions. Lewis had not only succeeded in leading the Corps of Discovery, also known as the Lewis and Clark Expedition, across the continent and back again, making him one of the most famous men in the country, but afterward,

Jefferson appointed him governor of the Louisiana Territory—a swath of wild country encompassing more than half of the young United States at the time. Jefferson and the entire nation were waiting eagerly for the young captain to edit and publish his expedition journals and maps, but Lewis seemed unable to focus or articulate what he'd experienced. Time was up though. President Madison had demanded his report, and Lewis could stall no longer.

What, if anything, Lewis saw in the Tennessee forest as he traveled can only be guessed. By all accounts he was drunk, irritable, and lost in a schizophrenic loneliness he couldn't claw his way out of. Some guessed his poor health was related to his inability to find a wife. He was good-looking, famous, and well-off, although he had money problems. Finding a wife should have been easy—unless, as many historians believe, he was a little off. There's talk of syphilis or malaria-induced madness, alcoholism and other addictions, not to mention groundless rumors he was gay. There's no question he was suffering from depression, anxiety, and excessive drinking. No one—at least no one who talked about it in public—really knew Lewis's inner world.

Lewis tried to kill himself twice en route to Washington—it's unclear exactly how. His companions disarmed him and placed him under watch. After days of erratic behavior while detoxing, he appeared to return to reason and continued on his journey. The night of October 10, Lewis rode ahead and took a room at Grinder's Inn on the Natchez Trace Trail. The inn was owned and operated by Priscilla Grinder and her husband, but the latter was away on business. Lewis claimed to have been unable to sleep in a bed since returning from the expedition, so he had his servant spread out a grizzly skin and his buffalo robe on his cabin's floor.

Perhaps he lay in the thick, wild-smelling fur for a while before his racing thoughts and despair became too much. He took to pacing the room and talking loudly, even yelling, to himself and walking circles around the pelts of the two animals that best symbolize the American West. What was the story behind each animal? Lewis and the expedition had killed so many bison that it was unlikely the robe had any more significance to the captain than warmth and fashion. The bear pelt, on the other hand, was different. For Lewis, no other animal in North America compared to the brown bear. It represented the most mysterious and dangerous elements of the wilderness of the New World. Even the woolly mammoth, which at the beginning of the expedition Jefferson and Lewis had speculated still existed in the North American wild, paled in comparison.

THE BROWN BEAR FIRST MIGRATED FROM ASIA ACROSS THE BERING LAND Bridge to North America around 100,000 years ago. There it met a host of ferocious predators like steppe lions, saber-toothed cats, dire wolves, and—kings over all—giant short-faced bears. With a fourteen-foot arm span, the short-faced bear weighed a ton and on all fours stood eye level with a grown man. Its skull looked like a cross between a bear's and tiger's. There are some opinions that the short-faced bear blocked brown bears from migrating south, and whether that's true or not, little fossil evidence shows brown bears in the contiguous United States until 13,000 years ago, when the short-faced bear and much of the continent's megafauna appeared to have died at the spears and wit of the Clovis people and environmental factors brought on by climate change.

The Clovis were some of the first people to migrate across Beringia, and their culture evolved to hunt—and survive—the megafauna of America. Their spears had heads between two and nine inches in length attached to wooden shafts and could be thrown, jabbed, and even braced against the ground to impale a charging animal. A spear may not seem like much when faced with the zoological reality of the time, but the Clovis people successfully spread across the continent hunting mammoths, mastodons, giant ground sloths, and a wide array of ungulates. Following along in the fossil record, not long after nearly all the megafauna vanished, the Clovis spearheads disappeared, replaced with weaponry suited for smaller and less fierce animals.

For 13,000 years, the versatile brown bear reigned supreme in the animal world. The bear's survival was a testament to its intelligence, omnivorous diet, and adaptability, including its ability to go into a torpor during months with low food availability. The brown bear's range extended from the Arctic Ocean south to northern Mexico and as far east as the Mississippi River. When the first brown bears migrated into northern Southeast Alaska around 12,000 years ago, they encountered an arctic environment as giant glaciers and sea ice melted away, revealing the ABC Islands of modern-day Southeast Alaska. There, it's theorized that brown bears met and interbred with a vanishing population of polar bears. To this day, the ABC bears' DNA is more like that of polar bears than other populations of brown bears, including mainland bears that live just four miles away across Stephens Passage, which separates Admiralty Island from the mainland of Southeast Alaska.

The Tlingit people likely arrived in Southeast Alaska not long after the first brown bears migrated into the country. People and bears ate the same food, traveled the same corridors, and lived in the same places. On occasion, they hunted, killed, and ate each other, though for the people, brown bears tended not to be as important a food source as other mammals like deer and seals. No other animal evoked more taboos. A hunt involved elaborate ceremonies and a great amount of respect to the bear to ensure the hunter's safety and success, and to placate the slain animal's spirit. The hunting party usually consisted of a small group of men who used dogs, bows and arrows, and spears. They targeted animals three years of age or younger because the meat was more tender and better tasting. Guns and other elements of Euro-American culture would do much to change the dynamic of that relationship, but even in a postcolonial present, many Alaska Natives' speech and attitude toward bears are instructed by deep ancestral teaching. They recognized that they share many physical and social similarities with bears; many tribes viewed the bear as if it had at one time been human. The bear walks flat-footed like a person; when it stands on its hind legs, it looks like a giant hairy human; a skinned carcass looks similar to a large muscular man. Some still refer to the brown bear in familial terms, calling the animal Grandfather or Grandmother.

The first reference to a European seeing a North American brown bear was in 1602, when the Spaniard Father Antonio de la Ascension, a chaplain and conquistador, reported watching bears feeding on a beached whale in what is now California. The next came in 1691, when Hudson's Bay Company explorer and trader Henry Kelsey wandered across the Great Plains of Canada by foot and canoe—the horse hadn't made it that far north yet. Kelsey was the first white person to witness bison and described seeing "silver-tipped" bears, grizzlies whose guard hairs—the long, coarse outer fur on an animal—were silver in coloration. Kelsey reported that two bears traveling together attacked him and a Native—likely either a Cree or an Assiniboine—companion. Despite being armed with only a clumsy flintlock musket, Kelsey killed both bears and earned the nickname Little Giant. When Kelsey killed another grizzly on a different occasion, his Native companions warned him not to keep the hide. "They said it was a god and they should starve," Kelsey wrote in his journals.

In 1790 Hudson's Bay writer and explorer Edward Umfreville was the first Euro-American to record an encounter using the name "grizzle bear," which he coined after the silver coloration of the bear's fur. A few years later, the explorer and fur trader Alexander Mackenzie wrote of encountering a "grisly bear." Whether that's a misspelling or refers to the fear the brown bear inspired is up for debate. Many people, including Theodore Roosevelt, would later argue that the proper spelling and meaning is *grisly*. During the Corps of Discovery expedition, Meriwether Lewis guessed that the brown, grizzly, grisly, silver, yellow, and white bear were all the same species.

It would take biologists quite a bit longer to come to this same conclusion. In the early 1900s, C. H. Merriam, who headed the Bureau of Biological Survey for twenty-five years and was commonly known as a splitter—a taxonomist who classifies species based on relatively minor characteristics—came up with nearly one hundred different species and subspecies of brown and grizzly bears. Today it's agreed that all of Merriam's supposed species of brown and grizzly bears are one and the same: *Ursus arctos*. That said, physical and social characteristics vary widely, influenced mostly by geography. For instance, a big brown bear on Kodiak Island or the Alaska Peninsula may weigh 1,500 pounds but, due to the abundant availability of food and the close proximity of many other bears, will likely be relatively easygoing. A big brown bear in the taiga of Canada might weigh 500 pounds but, because there is not much food and because a low bear population density means it won't be as socialized, will have a higher likelihood of being aggressive. In nomenclature *Ursus arctos* is used to refer to a brown bear that lives within 100 miles of the coast. Grizzly bears, their inland kin, are referred to as *Ursus arctos horribilis*. They were given this designation in 1815 by the naturalist George Ord, a man whose only experience with grizzlies came from reading Lewis and Clark's reports.

In 1803, as Meriwether Lewis and the expedition ascended the Missouri River, the young captain frequently referenced the possibility of encountering the "yellow" and "white" bear in his journal. The only bear that Lewis and his men were familiar with was the American black bear. Because the black bear is smaller and more timid than the brown bear, people often have a hard time taking it seriously. Everyone, even the most seasoned guide, deeply respects, and often fears, brown bears.

Lewis wrote in his journal during his first winter at Fort Mandan, on the Missouri River in the center of present-day North Dakota, of their Mandan hosts' attitude toward the grizzly bear:

> The Indians give a very formidable account of the strength and ferocity of this anamal, which they never dare to attack but in parties of six, eight or ten persons; and are even then frequently defeated with the loss of one or more of their party. . . . When the Indians are about to go in quest of the

white bear, previous to their departure, they paint themselves and perform all the supersticious rights commonly observed when they are about to make war uppon a neighbouring nation.

The following spring, not long after leaving Fort Mandan, Lewis encountered his first grizzly bears, *Ursus arctos horribilis* by modern designation—likely either two subadults traveling together or a female with a big cub. He wrote of what transpired:

I walked on shore with one man. about 8 A.M. we fell in with two brown or yellow bear; both of which we wounded; one of them made his escape, the other after my firing on him pursued me seventy or eighty yards, but fortunately had been so badly wounded that he was unable to pursue so closely as to prevent my charging my gun; we again repeated our fir and killed him. it was a male not fully grown, we estimated his weight at 300 lbs.

He went on to offer a rather strange and seemingly inaccurate description of the bear's testicles being encased in separate scrotum sacks. Every ensuing encounter with brown bears they recorded in their journals told a similar story: men shot bears, and then wounded bears tried to attack terrified men before eventually dying. The amount of lead a bear could take and keep coming both awed and horrified the men. Lewis wrote after a handful of life-or-death encounters, "I find the curiosity of our party is pretty well satisfied with respect to this anamal."

The most spectacular encounter Lewis had with a grizzly was an anomaly to the rest of the stories he related. It was June 14, 1805, and Lewis had become the first white man to discover the Great Falls of the Missouri River the previous day. Lewis had just shot a buffalo and was watching blood stream from its mouth and nostrils when a large grizzly snuck within "20 steps before I discovered him." The explorer, who hadn't yet reloaded his rifle and had no trees to climb, wrote of the events that followed:

there was no place by means of which I could conceal myself from this monster untill I could charge my rifle; in this situation I thought of retreating in a brisk walk as fast as he was advancing untill I could reach a tree about 300 yards below me, but I had no sooner terned myself about but he pitched at me, open mouthed and full speed, I ran about 80 yards and found he gained on me fast, I then run into the water the idea struk me to get into the water to such debth that I could stand and he would be obliged to swim, and that I could in that situation defend myself with my espontoon.

Lewis wrote how he dashed into water until he was waist-deep and then turned to face his pursuer. When the bear reached the water's edge, "he sudonly wheeled about as if frightened, declined to combat on such unequal grounds, and retreated with quite as great precipitation as he had just before pursued me."

A deep silence came over Lewis not long after the strange bear encounter. He and his men made it to the Pacific Ocean, which should have been cause for celebration, but what Lewis thought or felt is unknown. He made fewer journal entries, which were more terse— during the year the men spent getting back from the Pacific, it appears he hardly wrote anything. If he had, perhaps we'd have a better understanding of the nature of the darkness that would consume him that night in 1809 at Grinder's Inn. The grizzly pelt that lay on the floor of Lewis's room likely belonged to an animal he'd shot on the upper Missouri River. Had the animal come for him, blood gushing from its mouth, with teeth bared? Did Lewis stand over the breathless carcass feeling he'd conquered the embodiment of the wilderness? In that cabin at Grinder's Inn, did he remember the bear as he charged his brace of pistols and put a barrel to his own head?

The shot failed to kill him—which is hard to believe, considering how efficient a marksman he was. He placed the second pistol's barrel to his chest, pulled the trigger, and then struggled out the door. That night's progression of strange events continued as he called out for water and someone to heal his wounds. The only people said to have heard his pleas, Mrs. Grinder and her children, were too terrified to respond.

At first light Mrs. Grinder and other witnesses approached and slowly entered the room. They found Lewis with a razor in hand "busily engaged in cutting himself from head to foot." He asked for water and complained about his strength, as it made it difficult for him to die. Not long after sunrise, lying atop the blood-soaked bearskin, he breathed his last.

MORE THAN THIRTY YEARS AFTER I ENCOUNTERED THE BEAR CARCASS IN a salmon stream with my dad, I pushed through the forest fringe and stood on a well-worn bear trail at the base of a mountain on Admiralty Island. I led the way, followed by my older brother, Luke, as we hiked through the rainforest. I tried to stay alert, as it was mid-September when bears are at their most dangerous, but my thoughts kept drifting back to Meriwether Lewis. I wondered if he had been carrying any sort of written report detailing the Corps of Discovery expedition when he killed himself. The White House swept his death under the rug. Some people had difficulty accepting his suicide and put forward theories that he had been murdered. I couldn't get the image of Lewis mutilated, covered in blood, and lying on a grizzly skin out of my head.

When Luke and I reached the alpine, we split up to cover more ground. My freezer was full, but friends who had been too busy to do much hunting could use some venison. I was feeling so distracted that I decided to go exploring instead. I clambered up steep scree to the top of the mountain where there was almost zero chance of encountering a deer. It was a crisp, clear day, and when I reached the summit, a perfect 360-degree panorama of mountains, ocean, and glaciers surrounded me.

I was walking along when I placed a foot in a hole and stumbled. Before me stretched what we in Alaska call a grandfather trail—a place where multitudes of brown bears have stepped in the same tracks and worn holes several inches deep. These trails are all over brown bear country.

Some are used for a season; others are traveled for generations. Biologists aren't entirely sure why bears, generally being solitary animals, go out of their way to step in each other's tracks. Often, the males urinate on their own feet and grind their paws into the earth to better leave their mark. Sometimes these trails appear suddenly in seemingly random locations and then vanish just as inexplicably. Other times they're along major travel corridors. Many of the trails high in the mountains are ancient—some likely date back thousands of years, to when giant ice caps began rapidly melting away from Southeast Alaska.

I knelt and placed my hand above a hole that had been worn into the earth by hundreds, maybe thousands, of bears. I thought about how everything leaves a trail, whether it's imprinted in the land, in the narratives we tell, or even in our blood. Meriwether Lewis, with the stories he brought back from the West, set the foundation for America's relationship with brown bears. A trickle, then a flood, of pioneers followed his trail west, doing their best to eradicate the bear as they went.

As I stood near that hole, my skin crawled, and I got the feeling I was being watched. I pulled my hand away and scanned the area before walking along the edge of the trail until I could no longer distinguish it from the earth.

CHAPTER 2

Blood Trails

They numbered no more than a thousand, maybe two and their heyday lasted less than 20 years, but the mountain men left an indelible mark on American and world history. . . . In 1856 Antoine Robidoux could account for only three out of 300 from thirty years earlier. In Arizona during the 1820s, James Ohio Pattie recalled sixteen survived out of 160 in one year on the Gila River. Fatal quarrels with friends, thirst, starvation, storm, accidents and disease took their toll but Indians and grizzlies seems to have accounted for most to "go under" as they preferred to call death. Grizzlies ran in packs of fifty and sixty back then and had no fear of man. Trappers reported seeing as many as 220 in a day. Weighing 1,000 lbs. and able to run to speeds up to thirty-five mph. Even when taking in their proclivity to exaggerate, the numbers must have been high.

—Marshall Trimble, "Mountain Men"

In 1814, five years after Meriwether Lewis was buried at Grinder's Stand, the Corps of Discovery's journals were published. In it, people read an account of a western America filled with seemingly limitless game, adventure, and economic opportunity. Indigenous peoples were described as being mostly harmless, with the implication that the only thing contending with white men for the dominion of the continent was the grizzly bear. Readers were familiar with the black bear but viewed the animal mostly as a resource and an occasional nuisance. The grizzly, on the other hand, radiated danger and presented an existential threat to the expansion of the young nation.

For Jedediah Smith, the book of Lewis and Clark's journals was akin to the Bible. Legend has it that after a friend and mentor gifted him a copy at age fifteen, he carried it with him for the rest of his life. Smith was in his early twenties when he traveled west to St. Louis in 1822 looking for employment and adventure. The world was mad for beaver fur hats, and the market's supply depended on American Indian trappers. Alcohol, and a lot of it, was involved in bartering. In 1822 it became law that booze could no longer be used in trade with Natives, which in part led to the formation of free trappers—parties of largely white trappers that companies would outfit to attain beaver pelts. Smith, dreaming of his chance to follow in the footsteps of Lewis and Clark, signed on with one such group, the Rocky Mountain Fur Company, founded by William Henry Ashley and Andrew Henry. He was joined by a ragtag assemblage of St. Louis drunks and dreamers who agreed to ascend the Missouri River and spend between one and three years engaged in the fur trade. In the ranks were Hugh Glass, Jim Bridger, and others who would become legendary mountain men. They became known as Ashley's Hundred.

It's uncertain why the grizzly, despite occupying what is now the Lower 48 for at least 13,000 years, never made it east of the Mississippi River. Most of Ashley's men had no experience with the animal—they no doubt spoke obsessively about the bear, retelling stories from the journals of Lewis and Clark, as they slowly lined their way up the Missouri River. Some of the more seasoned men had their own tales, many of which were almost too wild to believe. There was more danger than just bears stirring in the wild country they were headed into though. Relations with Indigenous peoples were becoming tenser as epidemics of disease, alcohol, and increased contact disrupted their cultures.

After a winter spent building forts and trapping beaver, most of Ashley's men were camped on a sandbar above the Missouri River near a heavily fortified village belonging to the Arikara people. The trappers were hoping to trade for horses to continue west in search of better beaver country. On the morning of June 2, between six hundred and eight hundred warriors opened fire on them in a surprise attack. Hugh Glass was shot in the leg before he, along with most of the men, took cover behind the gunned-down bodies of twitching, bleeding horses. Ashley had anchored his boat out in the river, and men in skiffs tried to rescue their companions. According to legend, Jedidiah Smith, while providing rifle fire, was the last man to abandon his position on the beach. Between twelve and fifteen of Ashley's men died in what became known as the first Plains Indian War.

In August of 1823, Smith, Glass, and fifty other trappers returned to the Arikara village with the 6th Infantry and seven hundred Sioux warriors—the Sioux and Arikara had long been enemies. The Sioux charged ahead and were met by a comparable force of Arikara in the fields outside the village. One witness likened the fighting to a bunch of enraged bees. The Arikara warriors were retreating into the protection of their village's fort as the mountain men and army arrived. The Sioux began mutilating and dismembering bodies, taunting their enemies to come out and fight. Fear suddenly rippled through the warriors as a Sioux shaman appeared amidst the carnage, crawling on all fours. The man growled and snorted, mimicking the movements of a grizzly. The Sioux begged their white comrades to look away as the shaman sniffed an Arikara corpse. With his teeth, he began ripping the flesh from the chest of the man. One can imagine young Jedediah Smith, along with the other mountain men, staring in terror and wonder.

THE SMELL OF THE SMOKE FROM THE RAZED ARIKARA VILLAGE WAS STILL on Hugh Glass as he followed Andrew Henry on a five-hundred-mile overland journey to Fort Henry, where the Yellowstone River joined the Missouri River. Partway there, Glass disobeyed orders, broke formation, and was traveling alone just beyond the other men. A grizzly rose from the brush, let out a low woof, and eyed the lone man. Nearby, her scared, anxious cub bawled. Perhaps Glass fired first, or perhaps the mere sight of a man was enough to make the bear charge. The other trappers heard Glass scream and hurried through the woods. After a volley of shots, the bear collapsed dead atop the mountain man. Henry and his men rolled the bear off to find Glass a torn and gruesome mess. Death would be merciful and quick, they assumed. The trappers made camp while he wheezed terribly through his torn throat.

Despite his being the subject of numerous books and a blockbuster movie, not a lot is known about Glass before he joined Ashley's Hundred. A trapper who claimed to have been acquainted with him wrote that Glass was once a sailor and had been captured by pirates and forced into servitude. The story goes that Glass and another captive escaped a few years later, and the two wandered across Texas and Oklahoma before being captured by the Pawnee. The trapper related that though the companion was burned at the stake, Glass spent several years living with the Pawnee until he showed up in St. Louis around 1822.

Night fell, and Glass would not die. The Arikara were in the area and eager for revenge. Henry knew the longer they stayed put, the greater the risk of losing more men, or even the entire party. He offered a substantial amount of money to two men to keep a death vigil and bury Glass when the time came. John Fitzgerald and a nineteen-year-old, who some have theorized could have been Jim Bridger, volunteered. Henry and the rest of the men continued west, only to be attacked the following night. In the early dawn they left two of their men buried in shallow graves marked by crude crosses. It's not known how long Fitzgerald and the youth waited with Glass. It probably wasn't long, considering the likelihood of an Arikara attack. The two covered Glass with the skin of the grizzly as he lay wheezing, then took his rifle and possessions before heading toward Fort Henry.

Maybe it was the woodsy musk of fur or the smell of the rotting bear's blood mixed with his own that roused Glass. He probably wondered if he was dead as he stared up at the sky. Much of what happened next is uncertain, but it's likely Glass crawled over to the bloody, bloated bear carcass and rested his arm and face on her ribs. He studied the surrounding woods before biting and tearing away chunks of meat. He ate as much as he could, then began crawling toward Fort Kiowa. Historians estimate between two and three hundred miles separated him from the fort.

Meanwhile Jedediah Smith, who'd recently been promoted to captain, was leading a party west in search of better beaver country. He was going into terra incognita, places that even his heroes Lewis and Clark had never explored. The trip had not been easy; they had nearly died of thirst while traveling across what would soon be known as the Black Hills of South Dakota. Not long after, the mountain men were leading their horses through the brush above a creek bottom. A large grizzly appeared, then supposedly charged the line of panicking horses. Smith, with his rifle cocked, stepped out to meet the bear. A moment later the bear flattened him and engulfed his skull in its jaws, ripping his scalp and part of his face. Next, the bear bit into Smith's side, picked the mountain man up, and shook him. Smith broke several ribs, but a musket ball pouch and hunting knife absorbed the brunt of the bite. The bear was killed before it could do more damage.

The mangled captain's companions looked at him and were at a loss as to what to do. His scalp and ear were hanging from his skull, and the side of his face was badly torn. They discussed their options before addressing their wounded leader. Smith calmly asked if anyone had a needle and thread. The items were procured, but the men were reluctant to do what needed to be done next. Smith took charge and politely asked one man to do him the honor of sewing his ear and scalp back on his head. For the rest of his life, Smith would wear his hair long and combed over to hide his scars.

While Smith convalesced, Hugh Glass was crawling toward Fort Kiowa. He'd let maggots eat his rotting flesh to ward off infection and subsisted on rattlesnakes and carrion. Glass found Fort Kiowa empty but soon encountered and joined a boat manned by five French traders ascending the Missouri River. He wanted a reckoning with the men who'd deserted him badly enough that he was willing to return to Arikara territory. In an event of brutal irony, over the course of the upriver journey, Glass just missed John Fitzgerald, who floated down the Missouri with two other men in the cover of night. At one point, Glass—even though he had open wounds and likely broken bones yet to fuse back together—was put onshore to hunt meat. He was quickly attacked by a band of Arikara. Legend has it that a Mandan horseman plucked him up and whisked him to the safety of J. P. Tilton's fur trading post, just down the Missouri River from the site of Lewis and Clark's Fort Mandan. The French traders were slaughtered by Arikara not long after.

Traveling alone, Glass finally caught up with Andrew Henry and his fellow trappers at the end of December. He limped into Fort Henry looking more dead than alive and promptly forgave the young man who'd abandoned him, supposedly on account of his youth. Toward the end of winter, Glass headed south in search of Fitzgerald, traveling with four other trappers tasked with delivering a message to Ashley. On the Platte River, the party was attacked by a large group of Arikara. Glass hid in the brush, watching while two of his companions were hacked up. He snuck away in the dark of night and began another several-hundred-mile journey alone across the wilderness. This time he at least had a knife. Sometime in June, Glass made it to Fort Atkinson, in present-day Wisconsin, where he found Fitzgerald. It's unclear what happened during the encounter, but Glass didn't kill the man. Instead, he took back his rifle and headed down the Missouri to St. Louis.

Some would think that after what Smith and Glass had suffered, they would seek out an easier, safer livelihood. Something had gotten into their blood though. Maybe it was the mist on a creek at morning's first light and the sweet smell of beaver. Maybe it was the wild open of the country and the herds of thousands of buffalo. Maybe it was the grizzly. Maybe they'd seen too much to ever be able to return to civilization.

Smith made it to 1831. He likely roamed farther than any other mountain man. Many scholars credit the routes he mapped with setting the foundation for western expansion. Comanches attacked him while he was traveling alone in Kansas. I wonder what his assailants thought when they lifted his scalp and saw his scars. I imagine the warriors plundering Smith's outfit and finding the book of Lewis and Clark's journals. I imagine them tearing out pages and scanning words like animal tracks. I imagine those torn pages rustling in the wind around Smith's mutilated corpse long after the Comanches rode on.

Glass made it until 1833, when, after spending years trapping beaver in the Southwest, he returned to the upper Missouri. Early that winter, near Fort Cass—not far from where the city of Billings, Montana, now stands—he and two other trappers were ambushed by Arikara while checking their beaver sets. The war party left pieces of their scalps impaled on sharpened sticks above their naked and mutilated bodies.

I GREW UP LOVING TO HEAR ABOUT THE MOUNTAIN MEN AND WISHING I could have lived during their era. My dad would sing the *Davy Crockett: King of the Wild Frontier* Disney theme song to me most nights before bed. The most evocative line was the one about him killing a bear when he was only three. I, too, wanted to match myself against bears and mountains. Even as a young adult, I imagined that the lives of men like Hugh Glass and Jedediah Smith possessed a sort of meaning and freedom that evaded my own.

In the fall of 2003, fed up with civilization, I dropped out of college and decided to roam the mountains. I was two hundred years late, but Alaska still had wild country, where you could wander for days without seeing a road or another person. Once, while camped in the eastern Alaska Range, I woke unable to move. Something—my first thought was a bear—loomed above my tent. I imagined it collapsing and tearing the tent and dragging me into the night. Instead, a voice spoke in a language I'd never heard. Other voices chattered in. It was obvious they were discussing me. One voice, calmer than the rest, seemed to exert control over the others, and soon their discussion faded to silence. In the gray dawn, I emerged from my tent and studied a wall of rugged mountains. It was a dream, I told myself. For the first time in my life, I worried I was losing my mind.

Ten days later an old trapper who ran a trap line off the Yukon River picked me up while I was hitchhiking to Fairbanks. We talked about hunting, bears, and wolves.

"I remember one wolf," he said. "She was pure white, almost as if she glowed. She reminded me of those arctic wolves you see in *National Geographic* magazines—the ones that live on those islands off the Arctic coast. She hung around my line for a couple months. I tried every trick I know, but I couldn't catch her in a trap, and she never let me close enough to get a shot at her. One day I saw her out around five hundred yards across the river. Shot her. Found a good blood trail but never found her. That was the most beautiful I've seen."

In the ensuing silence, I studied the trapper out of the corner of my eye. I mentioned a golden-red grizzly I'd had a standoff with high in the mountains a few days prior. The bear had refused to move when I clambered over a knife-edge ridge and came upon it—I'd wondered if it had been eating a Dall sheep. A grizzly protecting a kill, similar to a mom with cubs, is more likely to be aggressive, even attack. The bear and I had stared at each other for a while before I slowly backed away.

"A couple from California got ate last week out on the Alaska Peninsula. It's all over the news. This is the most dangerous time of the year. You be careful out there," the trapper said.

He dropped me off at a Kentucky Fried Chicken in the industrial outskirts of Fairbanks. A short while later, I sat with my greasy Styrofoam thinking about being eaten and wondering how strong a hold I had on my sanity. I glanced around at other people, then outside at snow-covered buildings rising into the gray, wishing I was back in the mountains.

A FEW DAYS LATER, I HITCHHIKED OUT OF THE CITY HEADING FOR THE Hayes Range for a monthlong wander. A young woman with a cooing baby gave me a two-hour ride to Delta Junction. Then, after I spent a couple of hours pacing back and forth in the snow on the shoulder of the Richardson Highway, trying to stay warm, a man with a big red beard picked me up. I put my pack next to a tote filled with spruce boughs, traps, and moose bones, hide, and scraps in the back of his truck.

"Going to make some sets for coyotes. Not enough wolves around anymore to keep their numbers in check," the man said.

He told me the Hayes Range was crawling with grizzlies and they were mean but all should be in their dens, except for maybe a winter bear—an animal that doesn't den or emerges early because it doesn't have enough reserves. These bears are desperate and the most dangerous.

After he dropped me off, I shouldered my heavy pack and trudged off into the snowy forest, dragging a sled with additional gear behind me.

A week later I watched a family of beavers frantically adding willow branches to their food cache. The world was shrouded in turbulent gray clouds; a blizzard looked like it would soon set in. A herd of caribou plodded through deep snow along a distant rise. It was mid-November, temperatures were falling rapidly, and days were growing shorter. I had not seen so much as a bear track during the last several days. In fact, in the five weeks since I'd dropped out of school and been in the field, I'd seen only two grizzlies. Still, most nights I'd lie awake thinking about bears, listening to the different sounds, and often wonder if one was approaching in the darkness.

Near dusk, a beaver and its kit crawled up onto the shore. I was hoping to spend another three weeks out and had been rationing my food; I was hungry in a way most people wouldn't understand. I checked my .22 rifle and moved quickly toward them. I crawled through snow until I was just feet away. I aimed at the kit's head and then shifted to its parent. At the crack of my .22, the beaver's head fell to the side, and it collapsed atop the sloping frozen mud. I ran and tried to grab hold as the beaver flopped into the lake, splashing into the black water, but when I reached in, I came up with nothing. Heavy snow fell as I paced the shore until last light, feeling nauseated with guilt, hoping the beaver would float to the surface. Its mate swam back and forth searching long into the night. When I woke early the next morning, the lake had iced over and the mountains glowed silver in the moonlight beneath a blanket of fresh snow.

I was inexperienced with deep cold, and a few weeks later I developed frostbite. It spread like a burn across my fingers, nose, cheeks, and toes. My feet were the worst—I could not feel them other than as a slight, numbed pain. I snowshoed for fifty continuous hours, through darkness streaked with aurora, across the tundra, and then west along the Denali Highway. The dirt road is closed to traffic during the winter, but it offered easier travel on snowshoes. A group of forty or so caribou spooked and came charging out of the brush across the road. The snow was so deep that the animals breaking trail weren't going more than five miles per hour. The caribou in the back of the progression were going faster and half-trampling the animals in front of them. At the rear of the herd were two gray wolves that looked almost as large as the caribou. The two wolves jogged along, the one in the lead seemingly touching the hindquarters of the caribou last in line. The wolf turned, looked back at its companion, and grinned. A few moments later, the caribou and wolves disappeared into the brush. Across the road, they'd left a deep trench of a trail flecked with droplets of blood glowing red against the white of the snow.

When I made it back to Fairbanks, I had to go to the hospital. The doctor who did my triage told me I was probably going to lose half my right foot to gangrene—in the end I lost only a minimal amount of flesh and toenails. I thought about the mountain men during my convalescence. Some claim they were heroes who ventured into the unknown and blazed a path for civilization across the continent. Others say they were murderers, even the perpetrators of genocide. I told myself I was no longer envious of their brutal and lonesome lives.

Not long after, I encountered the only grizzly I've ever seen in the Lower 48. It was in October, in the Lewis and Clark National Forest in

Montana, a few hundred miles from where Hugh Glass was killed. The night before, at a tiny road-stop bar, an affable American Indian bartender had warned me about how many bears were in the area. In the morning, I was walking along a trail through the woods when, from around a corner, a silver-tipped grizzly came walking my way.

The mountain men are long gone. The American grizzly, besides tiny remnant populations in pockets of Montana, Idaho, and Wyoming, has vanished too, but as the bear and I stared at each other, I felt a timeless electricity course through me. The grizzly, instead of attacking or displaying even a minor sign of aggression, swung around and lumbered away.

CHAPTER 3

California

Sorry as we are at the extermination of the California grizzly, its passing was an inevitable and necessary accompaniment of human occupancy of the land. Although none of us now can ever see the live animal, we take pride that the bear which was so much a part of our early history, carries on as the state's emblematic representative.

—Tracy I. Storer and Lloyd P. Tevis Jr., *California Grizzly*

In 2010, while on a six-week trek across the Brooks Range, I followed the fresh tracks of a large grizzly up a box canyon not more than thirty yards wide. Except for a lost Nunamiut Eskimo man I'd found and walked with for a day and a half to the village of Anaktuvuk Pass, I'd been alone since I'd begun traveling west from the Dalton Highway two weeks prior. The wind was hard on my face, and heavy rain and the roar of a cascading stream muffled out all but my loudest warning calls. Hours wore on, and my nerves became increasingly frayed—at any moment I expected to run into the bear. Being stuck in a narrow gorge with any grizzly is dangerous, but my instinct, honed from solitude and numerous tense encounters with bears, told me this animal was particularly aggressive. The tracks of two wolves appeared. A short while later, strewn across the ground, lay a caribou calf they had killed that morning. The bear had claimed it from them. Blood and offal blackened the sand and gravel, but little flesh remained. I knelt, cupped the calf's face with my hand, and studied the black scree mountain slopes rising into dark clouds.

I climbed out of the gorge, hoping to sidehill and avoid an encounter, but I quickly came to a crumbly cliff, so I sat in the pouring rain on a ledge and fought something akin to panic roiling inside of me. An hour later, I calmed down enough to continue. When the bear finally appeared, it looked like a gothic monster in the rain and mist. I let out a breath as it lumbered across the slope above, unaware I was watching, rolling over giant boulders in search of marmots. I hiked well past dark before camping near a mountain pass. The sound of caribou snorting and the clacking of their hooves and tendons woke me numerous times during the night.

The following day, I walked with thousands of caribou traveling down the April Creek valley. They had given birth to their calves on the Arctic's coastal plain in June and were slowly migrating south to their wintering ground. I pushed through a wall of brush and came upon the carcass of a partly eaten bull caribou lying in a stagnant pool. The surrounding mud was beaten down with the tracks of a small grizzly. A moment later, a dark form rushed forward from the willows—an old female caribou, which stopped a few feet away and stared. Late that evening, I dropped my pack on a knoll and was about to make camp when I nearly stepped on the bloody quarter of a caribou that had been killed a few hours prior. Arctic grizzlies depend heavily on a vegetarian diet but supplement their intake with caribou, ground squirrels, and other sources of protein when they can. Bears often drag a dead animal into the brush and cover it with debris in what is called a cache. They're especially dangerous when guarding a kill. The bear was probably just yards away with the rest of the animal. I pulled out my pistol, shouldered my pack, and hiked through the darkness as startled caribou ran circles around me until I was far away.

The population of grizzlies in the Brooks Range, especially in areas like Gates of the Arctic National Park and Preserve, which is closed to hunting, is denser than many might believe. When I first ventured into the Arctic, I figured my chance of seeing a bear was almost zero. I didn't bother bringing a gun and only threw in a pepper spray as an afterthought. That trip ended with me almost getting knocked down by a grizzly.

In the morning, a blond grizzly appeared beneath a black tor—a thirty-foot high protrusion of ancient granite rising from the tundra—and quickly walked in my direction. I backtracked and, hoping the bear had not seen me, deviated sharply from its trajectory. Soon the bear appeared and ran across the tundra toward me. It paused at seventy yards and paralleled my movement for twenty minutes across a big plateau until it lost interest and climbed a desolate ridge high into the mountains.

That night I sat cross-legged, watching the last of the day's light pour through clouds onto the rolling mountains above the headwaters of the Alatna River. It's near here that the Nunamiut believe their ancestors were given the gift of life. In *The Nunamiut Eskimos*, Nicholas Gubser relates that a benevolent giant named Aiyagomahala created the Nunamiut and then showed them how to hunt and trade. Aiyagomahala taught ethics, kindness, and love and warned about the dangers of anger. Before vanishing, Aiyagomahala stuck a mitten into the ground, and mountains—known as the Arrigetch Peaks—formed in its place to remind the Nunamiut of their creator. I felt there was something special about this place. The world appeared motionless; even the scattered bands of caribou seemed frozen.

IN 1826, THOUSANDS OF MILES TO THE SOUTH, JEDEDIAH SMITH LED A party of trappers across the Mojave Desert and the Sierra Nevada into the territory of California. Smith was looking for new beaver country and found a wilderness just as wild as the Brooks Range. Two years later he became the first American to leave a record of an encounter with the California grizzly, a now extinct subspecies of the brown bear known as *Ursus arctos californicus*. The California grizzly is said to have been similar in size to the giants of the Kodiak Archipelago—imagine a large male standing eight to ten feet on his hind legs and weighing up to 1,600 pounds.

Smith's journal entry is terse, telling how he snuck up and shot a bear and watched it run off wounded. The party of beaver trappers entered the Sacramento Valley and found it thick with grizzlies. They wounded several with their light rifles before killing a large male. A short while later, a trapper trailing a wounded bear was severely mauled. A few days after that, Smith was nearly torn up by a grizzly for the second time—he'd shot a bear, and while he and two men were approaching the dead animal, a different bear charged. Smith jumped into a creek, and the bear pivoted and fixated on another trapper. The man jammed his rifle's bayonet into the animal's neck and it ran off, crashing through the brush, leaving a crimson trail in its wake.

Early California stories tell of brown bears mauling Indigenous people on a regular basis. These incidents usually involved people accidentally crowding bears, triggering defensive attacks. Unlike the tribes of the north and Great Plains, it appears the peoples living in what is now California almost never hunted grizzlies. Authors Tracy Storer and Lloyd Tevis Jr. wrote in *California Grizzly* that bears had dominion over the land: "Early in the Spanish period of California, a sure way to earn the gratitude of the natives was to destroy grizzlies." The bear was not only a threat to life but also a frightening spiritual presence. Bear men and women practiced dark magic and assumed the form of a bear to murder people. Storer and Tevis wrote that the Pomo people believed

> the ghosts of the wicked Indians had to stay behind in the bodies of miserable and tormented grizzlies, forever roaming the wilderness to be hated and loathed by all who saw them. . . . The Chumash, living near the "Valley of Bears," a land of long grassy swales studded with oaks, where grizzlies were unusually plentiful, evidently believed that all who died there became grizzlies.

John Bidwell, who in 1841 was one of the first emigrants on the California Trail and later served as a California senator, wrote that "the grizzly bear was looked upon by the valley Indians with superstitious awe, also by the coast Indians. They were said to be people, but very bad people, and I have known Indians to claim that some of the old men could go in the night and talk with bears."

Spanish settlers first built a mission in California in 1769. Initially ravaged by disease and hunger, they were able to survive by eating brown bear meat until a resupply ship carrying livestock arrived. Horses, cattle, sheep, and goats were set free, and their numbers soon exploded. With a new and ample food source in the free-range livestock, the grizzly population grew rapidly too. Hunting bears became a favorite pastime of the vaqueros. These bloodthirsty men of remarkable courage hunted the bears on horseback using lariats. They would bait a bear with a slaughtered mare or cow, wait for the animal to come feed during the night, and then give chase. By working in groups—usually between four and a dozen men—they'd lasso a bear's paws and hind feet, stretch the animal tight, and then dispatch it by strangling it with a lariat. Or the vaqueros might bind or cage the bear before transporting it to sell at a town or settlement. Often, after Sunday church services, Californians would gather for entertainment that included grizzly bear and Spanish bull fights. Bull and bear would be tethered together, and the bear would be harassed with spears and dogs. Often the bear would kill several bulls before being fatally gored. Despite human bloodlust, the grizzly population remained high during Spanish and later Mexican rule of California.

In 1848, at the end of the Mexican-American War, the United States claimed California. Gold was found in the Sierra, and a stampede of more than 300,000 prospectors journeyed to California by land and water. One of these men was a thirty-seven-year-old Massachusetts shoemaker named James Capen Adams. He invested all his savings—and likely his father's too—in footwear to sell to prospectors, then left his wife and children to follow the stampede west. A cholera epidemic was ravaging St. Louis when he arrived. While he was arranging to transport his merchandise to California, a steamboat caught flame, and the city burned. All of Adams's investment was lost. Soon after, he received word his father had killed himself. The cobbler wandered the charred ruins of St. Louis as wagons passed by carrying corpses—more than 5,000 died of cholera that summer. He was faced with a tough decision: go home ruined or go west to seek fortune. He packed what few belongings he had and began following the Santa Fe Trail west. He suffered hunger and cold, nearly dying twice from disease before arriving in Los Angeles sometime in the fall.

American settlers had come to California in droves and, being well armed and possessing zero tolerance for anything that contested their dominion over the land, set about systematically killing grizzlies. Adams made and lost fortunes prospecting, ranching, and leatherworking "until at last," he said, "in the fall of 1852, disgusted with the world and dissatisfied with myself, I abandoned all my schemes for the accumulation of wealth, turned my back upon the society of my fellows, and took the road toward the wildest and most unfrequented parts of the Sierra Nevada, resolved thenceforth to make the wilderness my home and wild beasts my companions."

Adams hadn't always been a cobbler. He'd once worked as a hunter, trapper, and trainer of wild animals. His career ended when he made a near-fatal mistake by turning his back on a Bengal tiger he'd been training, leaving him scarred and with a permanent limp. In the Sierra, he reverted to skills he'd developed in his youth and set off on expeditions deep into the wilderness, sometimes going as far as what is today Washington State and the western edge of the Rocky Mountains. He hunted, trapped, and developed a deep affinity for grizzly bears. It's unlikely that anyone in history has been mauled by bears more times than Adams—one encounter left him with a crushed skull and part of his brain exposed. After a short convalescence, he went back to the business of killing and capturing bears and whatever else walked on four legs.

There are other, mostly forgotten, bear men who fill the history of the West. Two things set Adams apart from his peers. First, Adams would capture cubs after he killed their mothers, and with a firm hand and deep knowledge of animal psychology, he turned several of these orphans into servants and even friends. He'd use the bears as pack animals to carry hides of other animals, often grizzlies. He'd hunt with them—at one point a bear he'd named Benjamin Franklin saved him by fighting off a mother grizzly he had wounded. After a few years of roaming, as civilization turned places like San Francisco from camps into bustling cities, he took on the moniker Grizzly Adams and cashed in on the menagerie business with his captive grizzlies as the star attractions.

Second, in 1859 Theodore H. Hittell, a San Francisco journalist fascinated by Adams, spent an extensive amount of time interviewing the mountain/circus man and writing a book about his adventures. That same year, Adams—sensing he was dying and perhaps missing his family—loaded a ship full of his grizzlies and other beasts and, like a modern Noah, set sail for the East Coast. When he arrived in New York City, he was greeted by thousands of curious spectators. He rode a brown bear through the streets. Following behind him, horses pulled wagons of dozens of caged grizzlies and other wild animals from the West. He dismounted in front of a giant canvas tent that P. T. Barnum, "the greatest showman that ever lived," had set up in advance. For the price of an admission ticket, any New Yorker could look into the eyes of a brown bear, hear stories of the West, and watch Adams wrestle his bears.

For the short remainder of his life, Adams toured New York and New England as part of Barnum's circus. Legend has it that his already ill health went into full decline after a monkey he was training bit the exposed tissue of his brain. A serious infection spread, but Adams refused to retire. He was bent on getting the full salary and bonus that Barnum had promised him if he finished his contract. So, night after night, the dying man pantomimed adventures of the Wild West and pretended to fight grizzly bears. He fulfilled his contract, gave his wife his money, and five days later at his home in Massachusetts, died in a bed he hadn't slept in for twelve years.

In the decades after Adams was laid to rest, more and more white settlers flooded California, and the grizzly receded deeper into the mountains. Most bears were shot on sight by miners and ranchers. Professional hunters pursued warier animals with hounds, traps, and poison. Most sources agree the last California grizzly was killed in 1922, after a rancher discovered some of his calves had been attacked by a bear. A trap was set—for grizzlies' remarkable intelligence, they are generally surprisingly easy to trap—and a few days later the bear was found with its forepaw pinched between steel jaws. The *Visalia Daily Times* reported: "Smarting with pain caused by the jaws of the trap, the animal was furious and presented a sinister front to the approaching hunters, but a well-directed shot from a heavy rifle ended his calf-stealing propensities."

MY PARENTS GREW UP IN THE SACRAMENTO VALLEY; THEIR SUBURBAN homes were built over the bones of bears. In 1953, when my dad was four, California designated its extinct grizzly as its official state animal. The bear already had the honor of cresting the state flag and seal. When my parents were in their early twenties, they moved to Alaska for the same reason Jedediah Smith and James Capen Adams had gone to California: they wanted adventure, wildness, and economic opportunities. When I was a kid and we navigated the suburban sprawl, interstates, and smog to visit my grandparents in the Sacramento Valley, my dad would talk about what California used to be. It was a paradise lost for my dad but the fulfillment of the American Dream for others. There are still pockets in Northern California, he would say, that are good country.

During one of my last visits to California, I went to see my mom's father, who'd lost his wife a few years prior. They had spent much of their marriage arguing, but now that she was absent, my grandfather missed her terribly. He didn't remember a single fight and spoke of her like she was more angel than woman. Both my dad's folks, whom my grandfather had been friends with, were gone too. My grandpa, normally a restless, fiery man, seemed calmly resigned. He loved animals and nature, so we took short walks along the bike paths above the Sacramento River and near Folsom Lake. This was the wilderness Jedediah Smith and Grizzly Adams had once wandered. Now it was covered in concrete and homes with barely enough space for a few jackrabbits. In the evenings we sat on his back patio reading and listening to music. Occasionally he'd talk about how he'd migrated to California from New Jersey, where there was no nature, so his three children could have better opportunities.

When I was a kid I believed that, despite the urbanization of the rest of the country, Alaska would somehow always stay wild. Now, as I watch my parents age and Alaska become more encroached upon, I'm left wondering about the future. There's pressure from Alaska's political leaders and mining interests to build a 220-mile road across the southern flanks of the Brooks Range. Those same leaders and the Forest Service want to clear-cut most of the last stands of Tongass old-growth forest. The Pebble Mine project is on the eve of being permitted and putting the entire Bristol Bay region and the world's biggest run of sockeye salmon at risk. I keep thinking how not so long ago, California was just as wild as the Brooks Range—and how once it's gone, it's gone forever.

CHAPTER 4

The Man Who Killed Bears

The most thrilling moments of an American hunter's life are those in which, with every sense on the alert, and with nerves strung to the highest point, he is following alone into the heart of its forest fastness the fresh and bloody foot-prints of an angered grisly; and no other triumph of American hunting can compare with the victory to be thus gained.

—Theodore Roosevelt, *Hunting the Grisly and Other Sketches*

In 1893, Frederick Jackson Turner, then an unknown assistant professor of history, presented the theory that American identity and democracy was defined by the frontier. Perhaps the most interesting aspect of his thesis was that he argued that when the frontier ultimately closed, innovation and democratic ideals would also disappear. Pioneers had settled most of the country. Bison, numbering more than 60 million a half century before, had been all but exterminated. There had been an estimated 50,000 to 100,000 grizzlies in the contiguous United States when Lewis and Clark led the Corps of Discovery west. Now the grizzly was approaching extinction. Turner's frontier theory provided a foundation for the nation to begin to reassess its relationship with wild lands and animals. In 1894, Turner received a letter from Theodore Roosevelt praising his ideas and stating that they would be incorporated in Roosevelt's third volume of *The Winning of the West*. Roosevelt, a force unlike anything this country had seen, was about to change history with his promotion of wild lands.

Roosevelt had been born with a sickly body, an iron will, limitless ambition, and a fascination with nature. He was a man of paradoxes: he loved wild nature, believed in man's dominion over it, yet advised restraint so that future generations could use and appreciate it. He overcame life adversities by "practicing fearlessness." He was the prototype of the hunter-conservationist, an example that many hunters today claim to mirror and measure themselves by. His life was a whirlwind—one day the assistant secretary of the Navy; the next a colonel in the Spanish-American War; later that year the governor of New York; two years later the vice president of the United States; then, after President William McKinley was assassinated by an anarchist in 1901, president.

In the first year of his presidency, Roosevelt suggested that Admiralty, Baranof, and Chichagof Islands be set aside as a brown bear preserve. The idea was nixed, but during the next seven years while in office, he protected more land than any other American in history. Alaska author Kim Heacox wrote in *John Muir and the Ice That Started a Fire* that from

> 1903 to 1908, President Roosevelt would protect 230 million acres of federal land (in 53 wildlife reserves, 16 national monuments, and 5 national parks), an area larger than California and Texas *combined*. He had the long view, the courage to take criticism, and the wisdom to hear sage whispers among the howling dogs.

In 1884 Roosevelt had gone west after losing his first wife and his mother to different causes on the same day and in the same house. He hoped to flee depression, reinvent himself, and experience the frontier before it was gone. That fall he saw his first grizzly and, a moment later, put a bullet through its brain.

Roosevelt wrote of the hunt in his classic essay "Old Ephraim," which was a popular nickname for grizzly bears during the nineteenth century. In the Bible, Ephraim was the son of Joseph, founder of one of the twelve tribes of Israel. What exactly inspired this nickname is unclear, although at the beginning of the eighteenth century, there was some speculation that the lost tribe of Israel was living in the American West. Regardless, "Old Ephraim" referenced a clear link to the animal's humanlike characteristics.

Old Ephraim is also the name given to the last known grizzly bear in . Utah, which roamed the Wasatch Range until 1923. That bear was a giant male, infamous for preying upon sheep and evading hunters and trappers. It took Frank Clark, part owner of Ward Clark Sheep Company, nine years to finally kill the bear. When Clark wrote about the experience, he described hearing a terrible anguished roaring late at night coming from near where he'd set a trap. How he'd caught this exceptionally trap-wise animal isn't clear. He followed the bear, which had been caught on the front paw and was dragging a fourteen-foot log, for several miles before coming upon it. He aimed his rifle then shot six of his seven shots into the bear. His description has become area lore:

> I could see he was badly hurt evry time he would breathe squirt blood from both nostrils so I got up clos and fired my last shot into head or neck and down he went. I sat down and watched his spirit depart from that great body and it seemed to take a long time but at last he raised his head just a mite gasped and was still. Was I happy no and if I had to do over I wouldn't do it.

Two nights before Roosevelt saw and killed his first grizzly, he'd been camped with companions in the Bighorn Mountains in southern Montana when they were visited by the grizzly. It grunted and let out a "roaring whine" just outside the light cast by a campfire. In the most vivid passage of "Old Ephraim," Roosevelt described finding the tracks of the bear before he hunted it down:

> That afternoon we again went out, and I shot a fine bull elk. I came home alone toward nightfall, walking through a reach of burnt forest, where there was nothing but charred tree-trunks and black mould. When nearly

through it I came across the huge, half-human footprints of a great grisly, which must have passed by within a few minutes. It gave me rather an eerie feeling in the silent, lonely woods, to see for the first time the unmistakable proofs that I was in the home of the mighty lord of the wilderness. I followed the tracks in the fading twilight until it became too dark to see them any longer, and then shouldered my rifle and walked back to camp.

Roosevelt spent a day and a morning tracking the bear, until the trail took him back to the remains of the elk he'd shot a few days prior. The bear had fed on the gut pile and crunched the leg bones to lick up marrow and, after racking dirt and debris on what was left of the elk to bury, was bedded nearby. Roosevelt quietly approached, his rifle to his shoulder. When Roosevelt was ten paces away, the great bear lifted its head and its "small, glittering, evil eyes" locked on Roosevelt's. The hunter sighted his rifle on the bear's skull, exhaled, and pulled the trigger. Roosevelt would go on to kill more grizzlies, but this was the largest bear he would ever encounter and he "felt not a little proud."

Once, Roosevelt famously refused to kill a bear. It was 1902, and the president was on a hunt for black bears in Mississippi. Three days had passed and he had yet to sight a bear. His guides became worried so they used dogs to run down a bear while the president was lunching, lassoed it and tied it to a tree, and then summoned Roosevelt to kill it. Roosevelt refused to do so and walked away in disgust—after which someone else was said to have made a mess of killing the bear. A short while later, in the *Washington Post*, a political cartoonist published a rendition of Roosevelt pardoning a bear as a guide held it with a rope. That image gave birth to the creation of the teddy bear. Unwittingly, Roosevelt had turned the bear into the country's most popular comfort object for children.

WHILE GROWING UP, I'D LISTEN TO MY DAD SPEAK REVERENTLY OF Theodore Roosevelt. We were a hunting family—the meat we ate came from the deer Dad hunted and the salmon he caught. Our bookshelves held a handful of books Roosevelt had written and biographies about him. I remember reading stories about him going on safari and killing multitudes of African wildlife, knifing a cougar to death, and lamenting the eradication of the American bison while he was hunting some of the last of their population. My dad would sometimes question Roosevelt's concept of manliness, pointing out his desire to engage in combat and kill another man, which he did in Cuba during the Spanish-American War. Then he would go on to praise Roosevelt by relating how after he wasn't reelected president for a third term, he went on a yearlong expedition through terra incognita in the Amazon that almost killed him.

"Can you imagine any other president doing that?" Dad said, shaking his head in awe.

My dad didn't share Roosevelt's love of sport hunting though. Dad's core philosophy, passed down to my brothers and me, was to hunt only what you eat. The Alaska Department of Fish and Game does not require a hunter to salvage brown bear meat because it's generally considered foul tasting and often has parasites. Instead, a successful hunter brings home the skull and hide. Very few people, even people living in the bush, hunt brown bear for food anymore. About two-thirds of brown bear hunters are nonresidents who, by law, have to hire a guide at prices ranging from $10,000 in an area such as the Arctic, where grizzlies are smaller and considered lower in trophy value, up to $35,000 for areas like Kodiak Island and the Alaska Peninsula, where bears grow the largest and, hence, are considered to have the greatest trophy value. For my dad, trophy value was not part of the hunt. He taught us to respect brown bears and also that killing one, except in self-defense, would come dangerously close to crossing a moral line.

I've never had the desire to kill a brown bear, but at a certain point I began to wonder if the only way to really understand Roosevelt and the psychology behind brown bear hunting was to hunt a bear myself. During the summer of 2018, I couldn't stop dreaming of being alone on Admiralty

Island following a bear trail up a salmon stream with the wind on my face and coming face-to-face with an old battle-scarred male—a bear's sense of smell is said to be four times better than a bloodhound's, but its vision and hearing are comparable to our own. Slowly, I'd bring my rifle to my shoulder, push off the safety, and aim.

One of the most bear-obsessed people I know shot a brown bear and then ate the entire animal. It was spring, before the bears had gotten into the salmon, and the animal was a four- or five-year-old subadult whose meat would have tasted better than that of a mature animal—a male doesn't reach maturity until around ten years of age and even then continues growing for several years afterward. My friend knew about the danger of contracting trichinosis but consumed much of the meat raw. When I pressed him on why, he grew uncomfortable and eventually admitted he'd done it to know the bear more deeply. Like my friend, I wondered if hunting a brown bear would deepen my understanding of the animal. Unlike him, I doubted my fortitude in eating the three to four hundred pounds of meat that an adult male bear would yield. Nor did I have any use, let alone space, for a skull and hide.

I considered going back to the area where a big bear had picked a fight with me and my older brother. I had a hunch the same bear may have mauled a man some years prior. Could I justify killing him if he came for me a second time? That bear was the sort that preyed heavily on cubs and smaller bears. I'd often heard it stated that bear hunting actually increases the bear population because it targets old males, who account for the high cub mortality rate. I considered making a hunt but, at the last moment, not pulling the trigger. That seemed almost as morally questionable as killing a bear. But I still wanted to understand what was behind hunting a brown bear. The bear hunting guides I knew all seemed to have a deep respect for both the animal and wild country. Most had never shot a brown bear for themselves and were ambivalent about the killing part of their job. One guide, Ed, tried to explain it to me, saying, "Brown bear hunting is intimate. They're the biggest land predator, and that excites some folks. They're not like any other animal you hunt. It's really intimate. I enjoy hunting them but don't like killing them. I wish more photographers were willing to sit in the rain and wind and suffer for a week."

After our talk, I was left thinking about how often Ed had used the word *intimacy*. I'd noticed a similar desire for intimacy in the clients I took out bear viewing. People often wanted to be very close to bears, even expressing disappointment if the day passed without having an encounter within thirty yards. Many clients told me they wanted to look in a bear's eyes. I wondered what they hoped to see but never had the courage to ask. Nor did I have it in me to tell them that most often a bear looks in your eyes because it's feeling aggressive. Brown bears don't want intimacy from us—they'd be happiest if they never had to see or smell our species.

Dick Peterson, a veteran bear hunting guide and woodsman, put his bear hunting philosophy simply, saying, "Bears just want to be left alone. I have an incredible amount of respect for them. Going forward with humility and wonder and trying to open my clients' eyes to their amazing ecosystem was my job. Combating the common misperception of brown bears as demon monsters became a priority to me. Identifying the right bear and getting as close as possible for a clean shot was imperative to me. The enormity of the deed once you walk up to a dead brownie is something to behold; the raw beauty of the animal, spirit drained, was always a bit overwhelming. My years of brown bear guiding hold a special place in my outdoor experience, for the time spent in such a wild, intact ecosystem and in the presence of such an awesome predator."

The reverence with which Peterson spoke of the bear and wild country was something I could understand. Was the desire to experience wild country an underlying motivation for someone who wanted to kill a bear? A few beers deep, one guide told me with a bluntness that surprised me, his clients "just want to shoot something and go home. Bear hunters are basically all assholes."

It made me think back to Norman Mailer's 1967 novel, *Why Are We in Vietnam?* I read the book when I was teenager, at first thinking it was about Vietnam. Instead it's a portrayal of toxic masculinity that follows a group of Texans who fly to the Brooks Range to hunt grizzlies and anything else that moves. To Mailer, brown bear hunting was an extension of the worst and most violent parts of masculinity.

Another guide who asked to remain nameless also had little nice to say about his clients. I asked how many of his hunters he likes, and without missing a beat he said, "33.33 percent," and mentioned how a high percentage are oilmen from southern Texas. Yet a different man, who began working as a guide as soon as he graduated high school, claimed to like basically all his clients. When we talked about how controversial the bear hunting industry is—a different guide felt he was treated like a Nazi when people found out how he made his living—he shrugged and grinned lightheartedly. He couldn't imagine doing anything other than hunting bears.

Scott Newman, a veteran Petersburg guide, knew from age six that he wanted to hunt bears and fly floatplanes—and that everything else in life was secondary. He emphasized the ancient psychological element of man versus nature as being a driving force behind brown bear hunting; bear hunting, like driving race cars, bungee jumping, and skydiving, makes people feel more alive, and even virile, in the most primordial, existential sense. "There's something in our psyche when you're talking about our relationship with man and beast," Newman said. "Not everyone has it; only a certain segment does. It's primarily male. It probably goes back to prehistoric times when a man had to prove himself. For a man to stand up to the biggest beast on Earth—do I have what it takes? Fundamentally, that's what it is. They're the biggest, baddest thing out there. The power. They don't take anything from anybody. There are definitely some bad bears out there. There are enough people who get chewed on, enough that get killed. Clearly the majority of the time they back down, they allow that to you. They could shred you. Unfortunately, I know what that feels like."

On the spring day Newman was mauled, he was guiding a client on the east side of Admiralty Island. They were within rifle range of a large male bear, but before the animal offered a good shot and Newman was in position, the client started shooting. He hit the bear in the groin, hind paw, and through the meat of the forearm. Newman had time for a shot, which appeared to hit vitals, before the bear ran into the woods. It was late, and Newman wanted to find the bear and skin it before dark. He was following the blood trail into the woods when the bushes exploded and the bear came at him. He shot the bear through the chest when it was four feet away. Then it was atop him, biting, shaking, and tearing—it was so violent that Newman "kind of turned away from it." The bear quit its attack and sat down next to Newman. "We literally looked eye to eye that far apart—eighteen inches—eye to eye. It was the most malicious, malevolent look I've ever seen in my life. It was pure hatred. It was like, 'I want to kill you, but I can't.' . . . The bear ended up dying on my boot," Newman said.

Complicating the morality of brown bear hunting is the seeming paradox of the hunter-conservationist. Paul Johnson, who has been guiding bear hunters for four decades in northern Southeast Alaska, pushed for more protection for the species early on in his career. That caught flak from old-timers who viewed the brown bear as vermin best shot on sight. Johnson and other bear hunting guides were at the forefront of the battle to conserve the species and the huge stretches of wild habitat they need. They pushed for smaller bag limits and more restrictions, which resulted in a Fish and Game mandate that throughout much of Alaska a hunter is allowed to kill only one brown bear every four years—rather than one or two a season, as the allowance had been when Johnson was growing up in Southeast Alaska. Johnson's friend Karl Lane, also a bear hunting guide, was instrumental in saving Admiralty Island from being clear-cut logged, which would have significantly damaged the bear population. When cattle interests wanted to eradicate a portion of Kodiak Island's brown bear population, it was the hunting guides who fought the hardest to keep it from happening. There are more brown bears in Alaska today than at any time in the last 150 years because the animal is viewed as a big game trophy. Johnson and many other guides are key stakeholders and are committed to keeping the bear and the wild country around for future generations to experience.

"I care about those critters in a big, big way," Johnson said. "Is it just because I make a living off them? I'm sure that plays a role. I want to see them for my kids and other people to keep this business going. There's something about the bears. Well, there's more than something. There's a lot of reasons for it. When you're hunting, you see things nobody else does, especially in the dusk. Once this fall, I had twenty bears out. Not at the same time, but during that evening in the same spot. That evening we didn't even come close to taking one. We had one that was decent sized, but I couldn't confirm that it was a boar because they have longer hair in the fall. My hunter wasn't disappointed that night. He was excited about what he saw. People say, 'You go kill the goldarn things.' I say, 'Yep, I do.' They're precious. They're the soul of the country out here. The excitement of a hunter getting a good bear. Can I explain that emotionally to somebody? No, but I can say when we take pictures, the excitement is the same. The level of intimacy is higher when you kill it. That intimacy with the bear varies with the hunter. What draws people to hunt a bear is all over the board, just like what draws people who want to take their picture. The smell. Bears have got this beautiful sweet smell. Some people might say they stink. They don't."

ON THE EVENING OF SEPTEMBER 14, 2018, THE DAY BEFORE THE FALL SEAson opened for brown bear, I sat high on a mountain on northern Admiralty Island looking for deer. The hilly mountains of Chichagof Island were superimposed against the giant, white, rugged peaks of the Fairweather Range. A good-sized bear appeared a few hundred yards below, its fur glowing in the low sunlight. He wasn't a giant, but he was an adult male. He stopped frequently to graze on berries as he slowly swaggered up the valley. I tested the wind, studied the contours of the mountain, and envisioned a stalk. I imagined closing the hundred yards until I would be within easy rifle range.

The bear paused and lifted his nose to scent the gentle breeze. I thought of the meaning of intimacy. How when I'd told my friend Forest Wagner, who'd been attacked and seriously injured by a brown bear, how often the word *intimacy* came up in my conversations with guides, he'd said, "Intimacy? Like how a rapist feels to its victim? I don't understand why anyone would want to kill a brown bear."

I thought of the Canadian explorer Henry Kelsey, who, after killing a grizzly, was warned by the Natives he was traveling with that he'd killed a god. When Roosevelt lined up to shoot his first grizzly, and every bear after, was he hoping to kill more than an animal? I imagined him standing over the dead beast in mute wonder at what he'd done. If he could kill "the mighty lord of the wilderness," maybe, for a moment, he believed he could conquer his own frailty, anguish, and even death.

I studied the bear through my scope for a few more moments and then laid my rifle atop the soft heather. The bear slowly moved up the mountain as the sun sank low on the horizon. It was time to move on. Bucks would soon be creeping out of the woods to feed on the last of the year's alpine bloom.

CHAPTER 5

Breath of Wilderness

When one is alone at night in the depths of these woods, the stillness is at once awful and sublime.

—John Muir, *John of the Mountains*

The day John Muir encountered his first grizzly, in 1868, he had set out into the Sierra Nevada of California specifically to observe bears. He had vowed the year before to devote his life to studying nature after he'd suffered an injury, while working in a factory, that left him temporarily blinded. He wrote how he rushed two different black bears, each in order to "study his gait in running," but both animals stood their ground, irritated by the man's strange behavior. Later that day, Muir encountered a small grizzly and immediately fell to the ground and hid behind a clump of brush. He wrote of what transpired next:

> But he had either heard my steps on the gravel or caught my scent, for he came straight toward me, stopping every rod or so to look and listen: and as I was afraid to be seen running, I crawled on my hands and knees a little way to one side and hid behind a libocedrus, hoping he would pass me unnoticed. He soon came up opposite me, and stood looking ahead, while I looked at him, peering past the bulging trunk of the tree. At last, turning his head, he caught sight of mine, stared sharply a minute or two, and then, with fine dignity, disappeared in a manzanita-covered earthquake talus.

During his first summer in the Sierra, Muir worked at a sawmill and as a sheepherder—he repeatedly referred to sheep as "hoofed locust" for the environmental degradation they caused. Grizzlies had been preying on livestock, particularly sheep, for the last century. They mostly came at night, killing and eating before retreating into deep brush and mountains where they were safer from people. Most herders were poor, inadequately armed, and powerless to do anything. Often, the owner of the livestock would hire a professional trapper or hunter, who would frequently kill numerous bears in the process of trying to find the one guilty of predation. Livestock and timber harvesting were the greatest causes of the decimation of the grizzly and Sierra wild country. Later, Muir would lead an army of preservationists against the two industries.

On occasion Muir would listen during the night as a bear attacked sheep he was responsible for. He wrote that while a bear generally killed only one or two animals, ten to fifteen other sheep would be trampled and suffocated by the rest of the flock. The experience of lying in the darkness with, at most, a canvas tent wall for a buffer, listening as a bear killed and devoured a sheep nearby would have been unsettling on several levels. Muir believed in a romantic vision of wilderness and nature—passed down by transcendentalists like Thoreau, Emerson, and Whitman—as a thing of harmony and beauty. The grizzly's predatory nature was hard to reconcile within this frame of thinking.

During the following decade, as Muir explored and worked odd jobs in the Sierra, he made a name for himself as a naturalist, geologist, and wilderness prophet. He popularized the idea that people needed to go back to nature for both their spiritual and psychological well-being. Too, he practiced what he preached and roamed far and wide, writing ecstatically about his experiences. On any page of any of his books, a reader will find quotes such as, "Oh, these vast, calm, measureless mountain days, days in whose light everything seems equally divine, opening a thousand windows to show us God."

Glaciers enthralled Muir, and there was no better place to study them than Southeast Alaska. In 1879, Muir bought a ticket on the SS *Dakota*, motored up the Inside Passage, and disembarked at Fort Wrangell, near the mouth of the Stikine River. Muir wrote that the Tlingit and white prospector village was the roughest and most inhospitable settlement he'd seen—no small statement considering he'd just spent years patronizing mining outposts in the Sierra.

The locals weren't sure what to make of him when he confessed he had no interest in gold and that, instead, he was there to study the marvels of nature's inner workings.

Throughout Southeast Alaska, Christian missionaries were in the process of usurping the shamans who, until then, had helped hold the fabric of Tlingit culture together. Among other things, the shamans were able to communicate with the spirits of the animal world, rescue the lost, change the weather, and battle witches. Muir was invited to a Tlingit potlatch honoring three visiting Presbyterian ministers and their wives. The Tlingit hosts treated Muir and the Presbyterians to dishes made from canned Western food because they were worried they would offend their guests if they served Native food. Afterward, Muir and the missionaries were entertained with songs and dances, some of which were inspired by different animals. There was a pause in the performances, and suddenly a bear came bounding into the longhouse. The guests were startled for a moment until they realized it was a man wearing a nicely fitted bearskin. The man perfectly mimicked the movements of a bear, going so far as to pretend to catch and eat a salmon carved out of wood. The Tlingits then made numerous speeches renouncing their beliefs. Muir wrote that the last speech, given by Chief Shakes, concluded with these words: "Dear Brothers and Sisters, we have been long, long in the dark. You have led us into strong guiding light and taught us the right way to live and the right way to die. I thank you for myself and all my people, and I give you my heart."

At the end of the potlatch, Muir was gifted a shaman's headdress. It's unlikely he kept it, as he was bound for the wilds of the Stikine River, the Cassiar Trail, and Dease Lake in Canada. What was once one of the most powerful and feared pieces of regalia in Tlingit culture had been reduced to a trinket. (A century later, thanks to a Tlingit revitalization movement, this is no longer the case.) If Muir reflected on this, he did not share it in his published writings. He and the missionaries were certain that Native beliefs—full of strange taboos and dangerous superstitions—were a thing of the past. Muir sensed he was on the precipice of creating a new religion, and it was to the mountains he would go to receive his testament.

IN MID-SEPTEMBER OF 2014, ON THE ONE HUNDREDTH ANNIVERSARY OF Muir's death and the fiftieth anniversary of the Wilderness Act, MC and I were driving in Canada near the Yukon's border with British Columbia. We picked up a Tlingit man hitchhiking in Teslin, eighty-some miles northwest of Dease Lake. We told the man our plan to float the Nisutlin River and then ride bicycles seventy miles back to where we would leave our car. He became alarmed and grew more anxious when we admitted we didn't have a gun. His behavior struck me as odd; given what was about to happen, it makes me wonder if he knew there was an old and desperate bear in the area.

It was dark when we dropped him off and turned onto the Canol Road. The dirt road had been intended to open oil development in the Northwest Territories, but the project was abandoned. A few miles later, our car's headlights illuminated a sign that read "Breath of Wilderness" in front of a well-kept cabin. The place was home to Claudia Huber and Matthias Liniger, who owned an eco-adventure tour company. I'd noticed it in the past during this drive and always felt the cheery sign a bit odd, even haunting, hemmed in by the surrounding black spruce forest. We drove, rarely going much faster than twenty miles per hour due to how rough the road was, until, tired, we pitched our tent in a pullout.

I boiled water on our camp stove, dumped it into a plastic water bottle, and passed it into the tent to MC to help her feet stay warm during the cold night. Moonlight illuminated a forest of autumn-yellowed willow and mountains glowing with a fresh dusting of snow. I stared up at the stars until MC told me to come to bed.

The following morning, we floated down the narrow, windy Nisutlin River through the lonesome taiga. We came around a bend and saw a young bull moose at the water's edge. The bull studied us, tatters of velvet hanging from his small red antlers, and ran into the woods. He crashed through the brush paralleling the river. After he had traveled a few hundred yards, he leapt back into the river and waded the current toward us. At five yards he appeared to be on the verge of trying to cut us off. I spoke sternly to him and pulled out a can of pepper spray as he stared at us with bulging eyes. He followed us for a while longer, running in erratic circles, before disappearing into the forest.

Twenty minutes later, movement in the shadowy forest caught my attention. For a few seconds the woods came alive with wolves. I counted at least a dozen in varying colors. They leapt over down trees and ran through brush in near silence. A minute after the pack had disappeared, a straggling half-grown gray pup, its tongue hanging out, came running along the riverbank. That evening, while we were setting up our tent, wolves began howling in the brush just a few dozen yards beyond our site. Soon, others answered on the opposite bank of the narrow river. For several minutes the two groups communicated. Their song, one of beauty and terror, reverberated in the forest, in the mountains, and in us. Slowly, as the wolves receded into the woods, their howls faded to silence. For a long while, MC and I sat by a small campfire and listened to the river as darkness eased onto the land.

A month later, a supposedly thirty-eight-year-old scrawny and starving grizzly broke through the window of Claudia Huber and Matthias Lininger's Breath of Wilderness cabin. They ran outside with their dog and took refuge in two different vehicles. The grizzly repeatedly jumped on the hood of one of the cars until Lininger was able to scare the bear away by honking the horn. Huber left the vehicle she was in and ran to join her husband. The bear charged back, attacked Huber, and dragged her off into the woods. Lininger ran into the cabin, grabbed a rifle, and shot the bear until it died. He rushed his wife to the health clinic in Teslin, but it was too late.

MUIR WOULD BECOME THE MOST POWERFUL WILDERNESS PROPAGANDIST in American history. Others before him, like Thoreau, had preached about the evils of civilization and the redemptive power of nature but failed to have a wide influence. Muir inspired thousands of people across the country who were living urban lives to buck the yoke of civilization and head to the mountains. Wilderness, now that it had been subdued, was no longer viewed as evil. Instead, for those wearied with their industrial lives, the wild, according to Muir, offered a chance for salvation.

Muir made his final voyage to Alaska in 1899. He was part of a two-month cruise, funded and led by railroad tycoon Edward Harriman, that included many of the nation's prominent scientists, scholars, and artists. These prestigious men had been selected and invited by C. H. Merriam, head of the Bureau of Biological Survey. The journey was called a scientific expedition, but it was centered around Harriman's desire to hunt a brown bear. Harriman had been ordered by his doctor to take a vacation—and nothing seemed more rejuvenating than exploring Alaska, climaxing with the killing of a bear. The brown bear population in Alaska then was much smaller than it is today due to commercial hide hunting, and the expedition sailed for quite some time without success. Finally, on Kodiak Island, some of the eleven men who'd been hired as hunters funneled a mom and her cub toward Harriman and a large entourage of armed men. The tycoon was credited with killing the mother, while the ship's captain was said to have killed the cub. The expedition sailed west to the coast of Siberia and then motored home.

They paused briefly at a Tlingit village in Cape Fox near the modern-day city of Ketchikan. The village's residents were out at summer fishing camps. Not seeing anyone, the men sawed down the village's totem poles and loaded them aboard the cruiser.

Even though Muir found the killing of brown bears and the theft of totem poles disagreeable, he and Harriman became lifelong friends. One man fought to celebrate and preserve wild places while the other's life purpose was building railroads, roads, mines, dams, and other alterations of the natural world. Part of Muir's charismatic genius was his ability to communicate with, befriend, and inspire everyone from warlike Tlingit chiefs to the most powerful businessmen and politicians in the country.

In 1903 Muir accompanied Theodore Roosevelt on a four-day journey into the Yosemite Valley of the Sierra. The two became fast friends, staying up late each night and conversing in the flickering light of a campfire. One witness reported that the two were constantly trying to talk over each other. Muir argued America needed to act fast to preserve the country's dwindling wild places. Roosevelt agreed but believed the expansion of the country and economic growth, including industries like mining, livestock, and logging, could coexist with the preservation of wilderness.

Their only real disagreement stemmed from an argument over sport hunting. Roosevelt believed it to be the "ideal training for manhood." He possessed a nostalgic view of wilderness, believing that through the conquest embodied in both sport hunting and expeditions of epic proportions, twentieth-century man could venture into wilderness and be redeemed. Muir, on the other hand, offered a vision of humanity's future relationship with wilderness. He preached distancing oneself from savage processes like hunting, which he called childish murder, and developing an aesthetic appreciation of nature. They put their differences aside and worked together to protect more federal land than anyone in history.

It was Muir's efforts and vision that led to the legal definition of *wilderness* used in the 1964 Wilderness Act. The bill, passed by Congress that year, protected 9.1 million acres of federal land. The tireless advocate

Howard Zahniser, of the Wilderness Society, wrote the bill and definition, which states, "A wilderness, in contrast with those areas where man and his own works dominate the landscape, is hereby recognized as an area where the earth and its community of life are untrammeled by man, where man himself is a visitor who does not remain." While the act was a great achievement in regard to the preservation of wild public lands, its definition of wilderness has proven problematic. The biggest issue is that the wilderness is home for Indigenous people—or at least was, until many were forced off their lands.

IN LATE OCTOBER OF 2014, I WAS HIKING ALONG AN ALPINE RIDGE ABOVE Juneau as a brisk wind, carrying waves of snow, howled off the icefield. I had learned about the Breath of Wilderness bear attack a few days prior and was hoping a long mountain walk would help shake off the morbid weight it had left me feeling. To the west, the mountains of Admiralty Island towered white and jagged. Most bears were already in their dens. I imagined them lying asleep in the darkness, the slow rising and falling of their hulks as they breathed. There's a saying that bears own the night. Not only do they tend to be more active during nocturnal hours, but they also prowl the dark recesses of our minds, the nooks and crannies where we try to suppress our most ancient terrors and fantasies.

I noticed the fresh tracks of a wolf and deviated to follow its trail. A dozen mountain goats lay or stood complacently on the southwest aspects of ridges and bowls. A flock of shimmering white rock ptarmigan whizzed by and vanished into the valley below. A raven landed on my head for a moment, then hovered a few yards off to the side, eyeing me sheepishly. The Tlingit say it was Raven who put the sun in the sky and taught humans how to live when they were new to the world. Raven was the trickster god of Southeast Alaska until Christianity came and taught that ravens were only birds. The raven rejoined its flock, and they bobbed and weaved on the wind with comical exuberance.

The wolf's tracks veered toward a glaciated valley. Huddled against the piercing wind, I scanned the ridge and surrounding slopes. The mountain goats seemed undisturbed, so perhaps the wolf had doubled back. A bald eagle, sometimes called the brother-in-law of Raven in Tlingit stories, flew past and floated wildly on the wind for a few moments before disappearing into the valley. Movement caught my eye and made me freeze. Wolves. A pack of five ran along the ramparts of a mountain. They were pale. The leader, perhaps the alpha female, was entirely white. Trailing closely behind, with a gray back and head but white everywhere else, was an exceptionally muscled wolf. They stopped and stared at me as I sat down in the howling wind. A few breaths later, they ran, crested a ridge, and vanished.

As I hiked down the mountain, I thought about John Muir and how he told of finding God in the wilderness. Thousands, maybe millions, of people—folks like Claudia Huber, Matthias Liniger, and even me—have followed in his wake, hoping that we too might find something to believe in. Heavy snow began to fall. A black bear paused from rooting around in a patch of brush to eye me warily. When I skirted around, it gave me a sideways glance before returning to digging. I made it to the tree line just as the last light of day faded. The trail was well-defined so I traveled without using my headlamp until a darkness in the shadows moved.

I froze for several moments before speaking gently.

"I'm here. I'm sorry," I said, waiting for the night to explode with a bear.

Slowly, as the trees whispered and moaned, I turned my headlamp on only to reveal a stump covered in bushes shuddering in the wind.

CHAPTER 6

Killer of Beasts

I have followed a big grizzly for three days at a time, snow from 3 to 12 feet deep, never had a coat on, I killed him and then I eat his meat. No other man will take such risks alone—not even a dozen men together. I have killed the largest and best tribes of animals. I have hunted them so close that it would take longer for 100 of them to accumulate again than it took me to kill 1000.

—Ben Lilly, in *The Ben Lilly Legend* by J. Frank Dobie

One day in the 1890s, the story goes, Ben Lilly left his family to go hunting and never returned home. Lilly considered it his duty to God and man to kill bears and mountain lions—he killed wolves begrudgingly, either because he didn't respect them or because they were harder to chase with dogs. Killing big predatory animals was rooted to America's national identity, dating back to the days when Cotton Mather preached a concept of wilderness in which the devil reigned and savages, beasts, and monsters stood in the way of settlers bringing God's light to the continent. Not only was wilderness evil—it was one's moral duty to conquer it. This belief resonated especially strong in Ben Lilly. He was a houndsman, a hermit, and the hardiest and most indefatigable woodsman Theodore Roosevelt said he had ever known. A pious man, no matter where he was or what he was doing, Lilly always rested on the Sabbath.

In 1907, Lilly was the lead huntsman during a vacation the president took to Mississippi to hunt black bear. The hunt, with all its pomp and accessories, was memorable but frustrating for Lilly. It took several days, but Roosevelt eventually killed a female bear. Not long after, with the South nearly void of bears and panthers, Lilly went west to continue his war on wildlife. He, and other men, wiped out any animal viewed as being at odds with the livestock industry. At times he sold meat and skins or guided rich hunters, but for the most part, he was employed by the government and cattle ranchers. Lilly—and other hunters and trappers—would bring in the scalps of the animals to government agents and collect a bounty. Even among the greatest and hardiest killers of beasts, Lilly stood out. He likely killed more mountain lions and bears, and extirpated their populations from a vaster region than any other man in the history of the United States.

There are numerous odd stories about Lilly—he was almost a folk hero but, unlike Davy Crockett, never enjoyed the attention and exploitation of a storyteller like Walt Disney. Lilly held the belief that you took on the characteristics of what you ate. He subsisted mostly on lion and bear meat, and could jump incredible distances and possessed supernatural strength. Rumors say Lilly always slept out, regardless of the season or the weather, without blankets or a tent. Once at a winter "camp," another hunter watched him strip naked and take a "bath" in the snow. Tireless on the trail, he would track a bear or lion for days without pausing to sleep, eat, or drink water. He spent an entire summer and fall shadowing a large grizzly near Taos, hoping that W. H. McFadden, an Oklahoman oilman, would come to New Mexico so he could guide him to hunt it down. Killing the Taos grizzly meant everything to Lilly, and he was not impressed that McFadden would rather take a vacation in Europe. He was famous for allegedly killing bears and lions with a "Lilly" knife—a big bowie knife he'd forged himself. Some accounts have him condemning his quarry to hell before engaging in mortal combat, in which he always came out the unscathed victor.

IN 1909, WHILE BEN LILLY WAS EXTERMINATING GRIZZLIES IN MEXICO'S Sierra Madre, Geronimo, the legendary Bedonkohe Apache, died a prisoner of war in Oklahoma. The renegade had been born near the headwaters of the Gila River in the mountains along the Arizona and New Mexico border. Five years later, the Mexican government, feeling powerless to protect many of its citizens from Apache and Comanche raids, put a bounty on Natives' scalps. Rogues, desperados, and madmen of many colors and creeds tried their hand at scalp hunting. For decades, Geronimo retaliated with a relentless guerilla war targeting Mexican and US civilians, until he surrendered for the last time in 1886. While living as a prisoner of war in Florida and Oklahoma, Geronimo, under guarded escort, was often shipped around as a tourist attraction—sometimes billed as "the worst Indian that ever lived." In 1905, Geronimo accepted an invitation to be part of Theodore Roosevelt's inaugural parade. Afterward, he was granted a few minutes to plead his case to be returned to his homeland. Roosevelt, worried Geronimo would return to raiding and murdering, or at the very least elicit violence from whites, refused his request.

The same year Geronimo died, Aldo Leopold went west to take a job with the Forest Service in the territories of New Mexico and Arizona, where memories of bloodshed, hatred, and fear were still alive in the desert, forest, and rugged mountains. Even though the Apaches had been confined to reservations, a few hiding out in Mexico's Sierra Madre still made occasional small, inconsequential raids along the Mexican-American border for the next decade and a half. With Indians no longer posing a real threat, the US government was turning to ridding the country of grizzlies, wolves, and mountain lions that—like Indians—were seen as a direct threat to civilization. Until they were exterminated, cattle would be preyed upon. Often, when ranchers complained of depredations by a particularly crafty wolf or bear, someone like Ben Lilly would be sent in on assignment. One part of Aldo Leopold's job was to help facilitate these predator control programs. Killing "varmints" was an industry with many similarities to the scalp hunting trade. Leopold would examine the scalps, or whatever part of the animal was required for the bounty, then pay the hunter or trapper the appropriate amount.

At first Leopold did not question this way of thinking. Wolves and bears did kill livestock, after all. Wouldn't no wolves or bears mean a better population of deer and elk for hunters too? Gradually, however, he realized how disrupted and sickly ecosystems become when they don't have large predators. Without predators, ungulates are left to over-browse their range, which leads to epidemics of starvation and disease. There are also subtler reasons—grizzlies, for instance, spread nutrients and seeds throughout the forest in their poop and by transporting fertilizers like salmon carcasses. Big predators, Leopold came to understand, are a vital part of the ecosystem and, in a way, keep things in check. In his book *A Sand County Almanac*, he writes of listening to the howling of a wolf and realizing how:

> Every living thing (and perhaps many a dead one as well) pays heed to that call. To the deer it is a reminder of the way of all flesh, to the pine a forecast of midnight scuffles and of blood upon the snow, to the coyote a promise of gleanings to come, to the cowman a threat of red ink at the bank, to the hunter a challenge of fang against bullet. Yet behind these obvious and immediate hopes and fears lies a deeper meaning, known only to the mountain itself. Only the mountain has lived long enough to listen objectively to the howl of a wolf.

He goes on to describe encountering a pack of wolves in the rimrock and how he and his companions opened up on the animals with their rifles. A wounded pup crawled off into the brush, but the mother lay crippled and fatally wounded.

> We reached the old wolf in time to watch a fierce green fire dying in her eyes. I realized then, and have known ever since, that there was something new to me in those eyes—something known only to her and to the mountain. I was young then, and full of trigger itch; I thought that because fewer wolves meant more deer, that no wolves would mean hunters' paradise.

But after seeing the green fire die, I sensed that neither wolf nor the mountain agreed with such a view.

Leopold would become the most influential champion of the environment since Roosevelt and Muir, though he offered a markedly different perspective on our relationship with nature. He was one of the first powerful people to publicly question the way we treat large predatory animals. Muir had distanced himself from bears and wolves. Roosevelt called wolves "the beast of waste and desolation" but wanted bears and cougars around, in large part so future generations of sportsmen could hunt them. While Roosevelt's philosophy was rooted in conquest and human use, and Muir was most interested in the spiritual and recreational aspects nature offered, Leopold recognized that animals and land have inherent worth beyond human desire or use. In order to have a healthy biotic community, Leopold realized, there need to be top-level predators like wolves and grizzlies. Otherwise, the ecological edifice will crumble.

Some call Leopold an ecologist philosopher. He loved the idea of wilderness, but he was more moderate in his vision than Roosevelt and Muir. He preached a land ethic that was both simple and revolutionary: in order for people to be healthy, we need to have a caring and direct relationship with nature. This relationship encompassed ethical hunting, farming, and stewardship—not the sort of behavior championed by Roosevelt that included traveling far and wide to hunt numerous animal species, many of which were not used as food; and not the sort of farming that involved grazing giant herds of cattle and sheep, which were the underlying cause of bison, grizzlies, and wolves being exterminated. Instead, Leopold wrote of the importance of hunting locally for food and how even a small piece of land could produce a sustainable amount of both physical and spiritual sustenance if properly cared for. Leopold, in his humble and articulate way, brought it to the public's attention that humans are part of nature, not apart from nature.

In 1922 Leopold proposed that the upper Gila River watershed be protected. It was around this time that Ben Lilly killed the last grizzly in the region. Two years later, the headwaters of the Gila River became recognized as the first wilderness area in the National Forest System and was set aside as a large swath of land that would be mostly protected from industrial development.

Certain essays written by Leopold, like the passing of the wolf and grizzly from the Southwest, resonate so strongly that they've become literary classics. Escudilla Mountain—which rises along the Arizona–New Mexico border not far from the birthplace of Geronimo—acts as the setting for one of Aldo Leopold's most impactful essays. He wrote of a giant grizzly, the last in the region, who hid out on the mountain. Each spring the bear would wander down into a valley and kill a cow before spending the rest of the year on the mountain. No one ever saw the bear, although his giant tracks terrified and intrigued the cowboys and hunters that came across them. Once, Leopold saw a cow killed by the bear. Its skull and neck had been smashed into pulp. Roads were built beneath Escudilla, bringing civilization, and a government trapper arrived to rid the mountain of the grizzly.

Ben Lilly killed numerous grizzlies on Escudilla, but it appears the man credited with killing the mountain's last grizzly was someone else. The trapper spent a month trying to get a sight on the ghost of a bear but never came close to seeing the animal. The bear was trap wise, and no foothold trap—no matter how enticingly set—could fool it. Finally, the government trapper set up bait tied to a set gun. Never having seen such a trap, the bear tripped the string and killed itself. The bear's pelt, because of its poor summer condition, was left to rot. Its skull was shipped to a museum and analyzed by biologists, one of whom was very likely C. H. Merriam—the great father of mammalogy—who catalogued the Escudilla bear in the book of dead grizzlies. In his essay, Leopold, questioning what "progress" truly means, offered his final thoughts on the Escudilla grizzly:

The bureau chief who sent the trapper was a biologist versed in the architecture of evolution, but he did not know that spires might be as important as cows. He did not foresee that within two decades the cow country would become tourist country, and as such have greater need of bears than of beefsteaks.

The Congressmen who voted money to clear the ranges of bears were the sons of pioneers. They acclaimed the superior virtues of the frontiersman, but they strove with might and main to make an end of the frontier.

We forest officers, who acquiesced in the extinguishment of the bear, knew a local rancher who had plowed up a dagger engraved with the name of one of Coronado's captains. We spoke harshly of the Spaniards who, in their zeal for gold and converts, had needlessly extinguished the native Indians. It did not occur to us that we, too, were the captains of an invasion too sure of its own righteousness.

In 1935, Aldo Leopold, Bob Marshall, and six other prominent conservationists created the Wilderness Society. A year later, Ben Lilly died believing he'd assured his place in heaven, at least in part for ridding the country of "varmints." During the last decade of his life, Lilly was obsessed with writing his autobiography. Words came harder than hunting, though, and a book never materialized. In his last years, while he was cared for in a home in Silver City, New Mexico, he found solace in making bizarre, childlike paintings of animals and pointing out mountain lions that only he could see. Some called him the last of the mountain men.

Despite Leopold's work, he lived to see the grizzly and wolf extirpated from state after state. There was an estimated population of a half million wolves in America when Lewis and Clark went west. When Leopold died in 1948, wolves had gone extinct in all but northern Minnesota and Alaska. Likewise, grizzly bears had gone extinct in all the contiguous US except for tiny populations in Montana, Idaho, and Colorado. Today all the grizzlies are gone from the Southwest. Black bears and mountain lions still roam in small numbers in a landscape increasingly fractured by development. There are rumors of a grizzly or two left in Mexico's Sierra Madre, hiding out much like the last of the Apache renegades.

WHEN I WAS SIXTEEN OR SEVENTEEN, I ENCOUNTERED A WOLF FOR THE first time. I was in the process of learning how to hunt, and even though killing a wolf was as taboo in my family as killing a brown bear, I was eager to prove myself. My dog and I were hunting grouse and ptarmigan in the mountains above Juneau when clouds descended and left us marooned in a whiteout. I noticed a shape moving through the swirling fog, and a wolf appeared. It looked young and emaciated even in its thickening winter pelage. The wolf came closer, sat, and stared at me and my dog. My dog quivered against my leg until the wolf shifted, and then he barked and leapt forward. I called him, and he hurried back to my side. A few minutes elapsed before the wolf rose to all fours and began nibbling at yellowed grass poking through a few inches of wet snow. Looking for a way down the mountain, I wandered in circles in the clouds for an hour— the wolf, all the while, trailing a distance behind—before I gave up and hunkered in the lee of the wind against a boulder. My dog shook as the wolf approached within forty yards, then sat and stared. I chambered a round in my .22 rifle, confident I could strike the wolf between the eyes and destroy enough of its brain so that it would die nearly instantaneously. I rested the rifle atop my backpack, holding the crosshairs on the center of the wolf's skull.

I knew a little about Muir and Roosevelt but another decade would go by before I would learn of Leopold. Still, the stories and ethics he offered had been absorbed by my dad and, through his teachings, passed down to me. I did not know it then, but my decision of whether or not to shoot the wolf was connected to the visions and teachings of all three.

In the two decades that have passed since, all I can remember of the animal is a shadowy form surrounded by swirling gray. Even the image of my dog, my best friend during my childhood, has blurred and faded. Oddly enough, I can still feel the rifle's stock against my cheek. I can still hear the soft but echoing click of the safety as I switched it to fire. I can still feel the touch of the trigger on my finger and the feverish buzz that came over me. I can still remember imagining my dad's disapproval at my choice to kill a wolf, especially one that followed me around like a lost puppy. I can still remember imagining how other men would praise me and think that since I'd killed a wolf, I was a great hunter. I can still remember an emotion that was something like shame hitting me and how hard it was to breathe as I unchambered my rifle and laid it down. Most clearly of all, though, I remember how the wolf sat motionless and watched me. How, while I wrestled with life and death, the wolf didn't flinch.

CHAPTER 7

Doonerak

Now we were back among people in Wiseman. In a day I should be in Fairbanks, in two more in Juneau, in a week in Seattle and the great, thumping, modern world. I should be living once more among the accumulated accomplishments of man. The world with its present population needs these accomplishments. It cannot live on wilderness, except incidentally and sporadically. Nevertheless, to four human beings, just back from the source streams of the Koyukuk, no comfort, no security, no invention, no brilliant thought which the modern world had to offer could provide half the elation of the days spent in the little-explored, uninhabited world of the arctic wilderness.

—Robert Marshall, *Alaska Wilderness*

In the 1930s Robert Marshall—widely known as Bob—spoke to Tobuk, an Iñupiaq man from the Kobuk River, who explained to him how the world works. Marshall summarized the concept in his book *Alaska Wilderness: Exploring the Central Brooks Range*:

Tobuk told me about the 'dooneraks' who were something like spirits, but a little less personal, and who were responsible for everything that transpired on earth. There were thousands and thousands of dooneraks, each with different ideas and objectives, often thoroughly antagonistic, and the happenings of the world represented the balance between their innumerable objectives.

Marshall was born in New York in 1901 to a wealthy and influential family. From early in life, beginning in the Adirondacks, he loved exploring wild places. After college, he went west seeking adventure, first in Montana and then, in 1929, Alaska. He sought out the Brooks Range, which at the time was a huge blank spot on the map, to live out a lifelong fantasy of making a wilderness journey akin to Lewis and Clark's. His first venture in the Arctic mountains left from the small mining community of Wiseman and headed into the upper reaches of the North Fork of the Koyukuk River. He coined the term "Gates of the Arctic" to describe the valley between Boreal Mountain and Frigid Crags, and it would later become the name for the second-largest national park in the United States. It was during this trip that he encountered his first grizzly bear while camped near the Continental Divide at the confluence of two creeks he named Grizzly Creek and Ernie Creek.

In August of 2017, I was camped at the exact spot where Marshall had his bear encounter nearly a century prior. My friend Ben Crozier sat staring at a camp stove, waiting for a pot of water to boil for dinner. Rain, snow, and wind had buffeted us since we'd walked away from Anaktuvuk Pass three days before. We'd seen four grizzlies and a wolverine during the trek. The first three had been a mom with two cubs born that year. We bumped into them at a dangerously close distance as they grazed on tart blueberries. Terrified, the mom ran up a hill until she realized her cubs couldn't keep up. She turned back, gathered her young, and ran up the Anaktuvuk River valley. Later that day, we watched a large male walk along the tundra a few hundred yards away. Oddly, a discarded bear spray with its safety off lay nearby.

The stove sputtered as I reread the passage in *Alaska Wilderness* about Marshall's first encounter with a grizzly. He'd had two packhorses tethered nearby when he noticed a large blond bear, followed by a smaller bear, striding across the mountainside toward the camp. The horses became agitated, and Marshall reached for his rifle. He wrote of what happened next:

I could not take aim at the bear without dropping the halter and losing Brownie [one of the horses], so I shot from my waist without aiming, still holding the halter rope tightly. I thought I would scare the bear, but the shot must have echoed, because the grizzly seemed to imagine it came from behind him. Anyway, he proceeded with double speed toward the tent. Now I knew there was no choice but to let Brownie go and shoot in earnest. I hit the bear, but not fatally and he turned around and retreated into the hills. The other bear had already disappeared.

The pot of water finally boiled, and we filled pouches of freeze-dried dinners with hot water. I placed mine inside my jacket to warm my cold stomach and then watched the last bit of swirling storm clouds dissipate beyond rugged black mountains. Marshall had named this area the Valley of Precipices. For him it had been the ultimate wilderness, a place that offered something akin to a religious pilgrimage and conquest. The Brooks Range fulfilled every aspect of the vision Marshall and the burgeoning wilderness movement were fighting for. Even today, despite significant oil development on the adjoining coastal plain, the Brooks Range is the wildest and least developed tract of land left in America. I looked down at Grizzly Creek and Ernie Creek, which were flooded and frothing. The last time I was here, Ernie Creek had been knee-deep and Grizzly Creek not much more than a trickle. I glanced south and saw two grizzlies rapidly coming across the tundra toward us.

"Bear," I said, alerting Ben.

The two were a blond mom and a big three-year-old cub. For a moment I wondered if the dooneraks were trying to tell me something. I began laughing, but the mom was coming in at a straight line, staring right at us, while her cub snorted and huffed anxiously. At around forty yards, we slowly rose. I drew my .44 Mag and spoke gently. A moment later the two bears turned and fled, running across the valley and up into the mountains until they vanished.

I can't say for sure what the mother bear's intentions were, whether she was feeling predatory or just curious. My guess is a bit of both. Like a person, a bear possesses reasoning skills and is receptive to communication, mostly through body language. A small movement—standing up, sitting down, taking a step forward or back—is often enough to placate a curious or agitated bear. Speaking gently is my next line of defense. Before I learned to talk to bears, I followed popular guidelines and yelled, banged pans, and waved my arms above my head when bears approached. I learned that if a bear didn't run initially, my yelling and exaggerated movements did little to deter the animal from coming nearer—and could even cause a bear surprised at a close distance to attack. Now I'm loud and aggressive only when a bear seems imminently about to make contact. I make noise, too, while traveling through bear country where my vision is impeded by things like brush, to give bears a chance to avoid a close encounter in the first place.

After our encounter with the two grizzlies, a deep silence came over the land. I listened for the different stories the Brooks Range had given me—the tundra pulsating, a deranged wolf howling, an old woman crying, the rush of a charging grizzly, the flesh and blood of caribou—but there was only the sound of my breathing and heartbeat. I scanned the dimming mountains before climbing into my sleeping bag and zipping the tent shut. That night as I listened to the raging creeks, I wondered about fate. I thought about the bear that had nearly knocked me down on my first trip into the Brooks Range—the encounter had happened thirty miles as the crow flies from where we were now camped. For months afterward I had something like PTSD (post-traumatic stress disorder) whenever I was traveling a willowy creek bottom. Now, thirteen years later, I wondered if the bear was still alive and if we might meet again. I thought of Mount Doonerak, a mountain that Marshall had named and obsessed over. He called it the Matterhorn of the Arctic, wrongly believing it to be over 10,000 feet tall and the highest mountain in the Brooks Range. He devoted much time and effort to trying to climb the 7,500-foot mountain.

Tomorrow, in the evening, Ben and I would reach the base of Doonerak. If the weather improved, we had enough food and time to spend three or four days trying to find a route to its summit.

SOME EXPLORERS HAVE CALLED THE BROOKS RANGE A LANDSCAPE OF anguish. During much of the year, it's locked in extreme subzero temperatures, storms, and darkness. On another trip Ben and I made in March, while cresting a mountain pass, I removed my snow googles to try to decipher a way through a blizzard and momentarily froze my eyes in their sockets. On the first day, after taking a shit, Ben had frozen every fingertip while wiping himself. The Nunamiut people, who've lived here for hundreds—maybe thousands—of years, call it a hungry country. They knew starvation well, as the land provided in fits and bursts of caribou and fish migrations and then lay still in desolation. Marshall, in contrast, extolled the beauty of the wilderness with a sort of levity and seeming naiveté.

After Marshall helped found the Wilderness Society in 1935, the organization quickly became one of the most powerful agencies promoting the preservation of wild lands. Marshall worked tirelessly to protect wild areas in Alaska and the American West from logging, road-building, and other commercial developments. Although the fate of the grizzly in the contiguous United States was pretty much sealed, Marshall and the Wilderness Society were instrumental in fighting for the remaining pockets of wild country where a few bears hung on. At just thirty-four years old, Marshall had become one of the most powerful voices promoting wilderness in American history. The *Jewish Tribune* called him the fourth most influential Jew in the world—no small thing to be lumped in with the likes of Albert Einstein. But in all his long hours spent alone in an office, rallying people, and socializing, Mount Doonerak haunted him. He'd already made two attempts on the mountain—expeditions that were months long—and had failed to reach the summit.

In 1939 Marshall returned to the central Brooks Range, hoping, again, to be the first person to summit Doonerak. His party made the long trek up the North Fork of the Koyukuk River to the base of the mountain. They made numerous attempts but were baffled by the maze of ridges, steep scree slopes, and cliffs. Winter arrives early in the Arctic, so Marshall soon retreated back to Wiseman and civilization. He wrote of accepting defeat in the final pages of *Alaska Wilderness*:

> One in a million, perhaps, could be a Nobel Prize winner or a President of the United States. The other 999,999 might burden their lives in gnashing their teeth over unrealized ambitions for greatness, or they might adjust to limitations and fate and get the greatest possible happiness out of the North Dooneraks, the Amawks, and the Apoons [smaller mountains around Mount Doonerak that Marshall climbed] which they could attain. Perhaps this philosophizing on a windswept pinnacle of rock might seem a little forced, but I could not help it, because I had talked only recently with an assistant manager, an associate professor, and a division chief whose lives for several years had been unhappy because they had not been promoted to head manager, full professor, and bureau chief.

A few months later, he sat on a midnight train from Washington, DC, to New York City. The train was chugging north through the darkness when his heart gave out at the young age of thirty-eight.

On schedule, Ben and I reached the North Fork of the Koyukuk in the evening and made camp above the flooding river. Something akin to grief had come over me while I traveled the same country, climbed the same peaks, and camped in the same places Marshall once had. I walked through thick brush and, thinking it might be nice to change up our freeze-dried diet, threw a line into a slough. A grayling struck my lure, and I brought it in. I studied the iridescent scales, oversized dorsal fin, and ancient eyes of the fish. Right then, the thought of killing the animal made me cringe. Gently, I freed the barb-less hook from its mouth and then watched it dart back into the pool. I went back to camp, sat close to a fire, and ate a pouch of freeze-dried chili mac with beef.

We spent two days hiking smaller mountains and waiting for the weather to clear. On the second day, as we neared a long ridge, we watched a dark grizzly grazing on berries. A few hours later, we sat near the summit as clouds swirled in the brisk wind, occasionally revealing the snowy ramparts of Doonerak across the Koyukuk River valley. The following morning, Doonerak still hidden in clouds, we inflated our packrafts, and the swift current of the river carried us out of the Brooks Range and into the lonesome taiga.

CHAPTER 8
Alaska Bear Wars

In 1908 Alaska's Governor Walter Clark requested a 12-month open season on brown bears. "They are dangerous animals," he explained. In 1915, Governor John Strong claimed that unless brown bears were exterminated, the raising of sheep and cattle in the territory would have to be abandoned. In 1919 Governor Thomas Riggs said, "The brown bears have no place in the economic development of the territory any more than the herds of wild buffalo would have in the wheat fields of Minnesota and the Dakotas."

—Jim Rearden, *Tales of Alaska's Big Bears*

In 1929, the same year Bob Marshall saw and then shot his first grizzly, a similar encounter occurred on Admiralty Island with a Forest Service employee named Jack Thayer. Marshall walked away from his encounter, but Thayer was killed by the bear after firing at and wounding the animal. Until 2018, Thayer's death was the only documented case of a bear killing a human on Admiralty Island. Growing up, I read short accounts of the Thayer mauling. Most intriguing to me was the story of a bear hunter named Hosea Sarber, who supposedly tracked the Admiralty Island mankiller down and then years later disappeared in the wilderness under mysterious circumstances. I read one speculation that Sarber wandered into the woods of Admiralty Island, leaving his gun behind, and was gotten by a bear.

In 2015 I wrote a column for the *Juneau Empire* newspaper about Thayer and Sarber and asked readers who had any additional information to contact me. A short while later I received an email from Betty Miller, an eighty-year-old woman who'd lived in Juneau her entire life, saying she was the granddaughter of Captain Carl Collen, who'd worked with Thayer, and that she had the original reports that Collen and Fred Herring, Thayer's assistant, had given to the Forest Service just days after the attack. She asked that I come look over the documents.

It was a dark, nasty day as I drove along Gastineau Channel. The ocean was frothing, trees and road signs were swaying in the wind, and rain was blowing sideways. Betty and her husband ushered me into their condo and sat me down with a cup of coffee at their dinner table.

"It was so terrible. If it had happened today, Jack would have been rescued and lived," Betty said as she passed me pictures yellowed with age.

Most of the images depicted the M/V *Weepoose*—used by the Navy during World War I and bought by the Department of Agriculture after the armistice—which Betty's grandfather had captained. In one image the *Weepoose* is tied up to a raft of giant logs. A young man and woman, perhaps Betty's grandfather and grandmother, stand on the logs, looking tired but happy, their hands interlocked and raised above their heads. I photographed all the pictures and documents, then thanked the Millers, drove home, and stared out at the ocean churning beneath the lights of the boat harbor. Rain drummed atop my roof, and the wind shook my home's walls. I spent much of the evening trying to get the image of Thayer, torn and lying alone in the rain and the darkness, out of my head.

ON OCTOBER 16, 1929, JACK THAYER AND FRED HERRING LEFT THE M/V *Weepoose* at first light and hiked up the valley behind Eliza Harbor on Admiralty Island. Thayer had been cruising timber in Southeast Alaska for the Forest Service since 1923; Herring had been his assistant for the past two years. People were looking for ways to diversify the region's economy as most of the giant mines were closing down and the seemingly endless supply of salmon was declining. Logging seemed to be the solution, and Admiralty Island became viewed as a valuable pulp resource.

Before a timber sale could be made, timber cruisers like Thayer and Herring had to inventory the rainforest. They would walk grids, measure tree diameters, and come up with estimates of board feet. During 1929, Thayer and Herring were working on one of the biggest pulpwood sales the Forest Service had ever proposed in Alaska. The sixty-foot M/V *Weepoose*, captained by Carl Collen, acted as Thayer and Herring's home base. Thayer, Herring, and Collen were at the end of several months of fieldwork and no doubt looking forward to returning to their homes.

Late in the rainy afternoon, Fred Herring staggered out of the rainforest, squinted through the gloom, and studied the high yellow ryegrass of an estuary. Bald eagles stared mutely from treetops, and ravens winged their way up the valley toward where he'd left his partner, Jack Thayer, mortally wounded, lying beneath dripping tree boughs. Herring looked back to make sure he wasn't being followed. A flock of Bonaparte's gulls called out raucously at the mouth of the stream. He followed game trails until he saw Carl Collen standing on the tidal flat waiting.

"Bear got Jack, but he is still alive," Herring told Collen.

In a statement made a few days after, Fred Herring wrote about the lead-up to the attack. Coming to a swampy area, they had noticed a bear tree—a tree that bears use as a marking and scratching post. Often called rub trees, these are all over grizzly country.

> We remarked on the size of the bear and the freshness of the chewing and clawing marks. We had seen many bear trees during the summer and so attached no importance to it.
>
> As we left the muskeg and entered the scrub timber I, who was in the rear about 3 or 4 feet, heard a snort and saw something move behind a clump of bushes about 15 or 20 feet behind me and to the left of our line of travel. I called Thayer's attention saying: "I think there is a bear or a deer behind that clump of brush, Jack." He and I stopped and watched and saw the bear raise up—his head and fore parts being plainly seen. He was a large brown bear very dark in color and apparently fully mature. I

said: "There's a bear, Jack" and I started immediately to run for a tree, as that is man's only refuge—the brown bear being too large to climb small trees. As I passed Jack he shot and almost immediately the bear began to bawl. I ran about 25 yards and climbed a tree which had limbs close to the ground and would afford speedy ascent. From the tree I heard the noise of a struggle and saw movements through the underbrush and then first realized what happened.

After a few minutes I saw Jack get to his hands and knees but fall again. I descended the tree and crept close with caution not knowing if the bear had gone far away. This was a short time afterwards, probably less than 5 minutes after I had climbed the tree.

Jack was conscious and said: "Where did he go, Fred?" I answered I did not know. He then said, "Save yourself, Fred," and lapsed into unconsciousness. . . . I removed my pack and shirt and laid his head on the pack with the large wound on his head up and bound my shirt around his head to hold the wound closed. I did what I could to make him comfortable and left for the beach where I met Capt. Carl Collen.

The two men rowed out to the *Weepoose* and threw together a first-aid and basic survival kit. They paddled back to shore, dragged the small raft above high tide line, and hurried into the dark and sodden rainforest. It was late, and besides the wind and dripping rain falling through a canopy of boughs and branches, it was eerily quiet. They followed a creek winding up a valley into the mountains. It was nearly dark when they found Thayer. They built a fire, dressed Thayer's wounds, and did their best to keep him warm, but he passed during the night.

I looked up from my photos of the reports as rain lashed the window and the darkness beyond, and thought of Betty saying, "If it had happened today, Jack would have been rescued and lived." Perhaps if it had happened today, he wouldn't have been attacked at all. At the time, the popular mind-set was that climbing a tree or shooting were a person's only defenses during a close encounter with a brown bear. Yet these were the two reasons the bear attacked. Running from a bear, which can be interpreted as prey behavior during most encounters, is one of the worst things someone can do. If the two men had stood their ground and talked to the bear, the chances are very good it would have let them be. The bear stood up not out of aggression but out of alarm to get a better look. When Herring ran, he heightened the bear's agitation, but even then it did not attack. Thayer was supposedly using hard-nosed bullets issued by the Forest Service, which are good for target practice but lack the mushrooming effect that is key to knocking down big animals. He'd heard all the stories about how grizzlies are bloodthirsty killers and probably figured if he didn't shoot the bear, he'd be mauled. He aimed, but in all probability his hands were shaking wildly. At the sound of the shot and the sting of the grazing bullet, a switch went off in the bear, and it retaliated.

Frank Dufresne, the Alaska Game commissioner at the time, attempted to track the bear down a few days later. He wrote in his book *No Room for Bears*, "An Indian tracker picked up the footprints and we followed them into the high meadows above timberline before they faded out. There was no sign of blood." Reflecting on the young forester's death, Dufresne continued:

> His death was to start another public outcry to remove all protection on Alaskan bears, put a bounty on the killers, poison them, get rid of them all. Of the several fatal attacks and maimings of humans by grizzlies, the case of the forest ranger was to create the most attention.

Six days after Thayer died, an editorial ran in the *Daily Alaska Empire* newspaper:

> Jack Thayer was a large asset in the development of Alaska. He was of the sort of young men that must carry into execution the plans of those who would make Alaska grow and prosper. He was the pride of his parents and idol of a charming girl to whom he was engaged. Yet if he were

the only man that brown bears had killed, we should say it was simply too bad. If there was not a constant fear that others will go the way he went, we should not ask that brown bears be exterminated in the parts of Alaska where men are engaged in the industries and settlement of the country. The brown bears serve no good purpose. They are essentially killers. It is bad enough to have them destroy game animals and salmon. It is too much when they turn their attention to people upon whom we must depend for [t]he development of Alaska.

It may be that there are places in Alaska where bears ought to be protected for the pleasure of sportsmen. Whether that is so or not, Southeast Alaska is not one of those places. We have too much to do to make room for settlers and industrial workers to permit this menace to hang over us. The brown bear in the First Division ought to be exterminated—and the extermination work ought to begin at once.

Many Alaskans believed you could have either economic development or brown bears, but you most certainly couldn't have both. Jay Williams, a timber cruiser and the Forest Service's designated bear expert, proposed exterminating all the bears on Admiralty to make logging the island easier. During this same time, livestock interests pushed to exterminate brown bears on a portion of the Kodiak Archipelago. Thayer's death was convenient to them, and they used it to bolster anti-bear propaganda across the territory. The Alaska Game Commission bowed to the wishes of political and industrial leaders and took away the majority of the already liberal regulations surrounding brown bear hunting.

The call for the widespread killing of bears backfired though. Thayer's death, and the push to eradicate Alaska's brown bears, would lead to the Save-the-Bear Campaign. Harry McGuire, the editor of *Outdoor Life* at the time, penned an extensive editorial in defense of the brown bear. In 1930 John M. Holzworth published *The Wild Grizzlies of Alaska*, mostly about the bears of Admiralty Island. Arthur Newton Pack, founder of the American Nature Association and *Nature Magazine*, published a feature about Admiralty and its bears in 1932. Both Holzworth and Pack were guided by a recluse bear hunter named Allen Hasselborg, who would be exploited as a pivotal force in encouraging America to reassess its relationship with brown bears. Sportsmen and other conservationists across the country rallied in support of the bear. With the grizzly already gone from most of the contiguous United States, suddenly Theodore Roosevelt's idea of setting Admiralty and the other ABC Islands aside as a brown bear preserve didn't sound so crazy.

NEARLY A YEAR TO THE DAY AFTER THAYER'S DEATH, HOSEA SARBER VIS-ited Eliza Harbor to hunt bears. Sarber is featured in many of Frank Dufresne's stories, as well as numerous other hunting articles and books. Dufresne was in awe of Sarber, calling his friend "the greatest bear student I ever knew." Sarber was born in Indiana in 1897, when thousands of hopefuls were stampeding north in search of gold. He was blinded in one eye as a small child but would later be renowned for his marksmanship despite this handicap. He came to Alaska as a young man, inspired not by dreams of gold or striking it rich in some other resource-based industry but by the desire to hunt. He became a well-known bear guide and game warden, and some say he killed more bears than anyone else in Alaska. Sarber wrote of what transpired during his bear hunt in Eliza Harbor: "I was wading down an Alaskan salmon stream at sunset when just inside the heavy jungle of alders and devils-club there came a sudden, resounding roar. Its volume echoed through the heavy timber like the voice of some mad creature out of this world."

As Sarber scrambled to the far shore of the stream, the roars continued, chilling him:

> I should have been pleased at the incident because I had been looking for a brown bear trophy all day. Standing my ground, I waited until the infuriated bear blew off steam. The loud roars gave way to huffing and blowing

as the animal backed away slowly. As soon as I was sure it had moved outside the stream-side brush into the open timber, I eased after it. It wasn't the smartest thing in the world to do, but I wanted that dark hide even though I realized the brownie might rush me at any moment. So it was with genuine relief that I finally spotted the animal. Carefully stepping from the thick brush into open timber, I saw the bear about forty yards away, standing with its back against a windfall, eyeing me with such burning hatred as I never hope to see again in any animal. Its wide head was swinging low from side to side and from its frothing lips came a sort of moaning and coughing sound. Here was a bear in the very instant before its charge!

I brought up my rifle, trying desperately to keep from shaking. The gun barked and the hand-loaded, 200 grain '06 bullet, pushed by a full charge of powder, felled the bear in its tracks. It died instantly.

Sarber found a scar on the bear's chest, where he believed a bullet had brushed its ribs. Imagining how the bear that killed Thayer had stood straight up, right in front of the man, Sarber theorized this was the same bear. Dufresne wrote, weaving in quotes from Hosea:

Hosea wasn't proud of the achievement. "That grizzly knew I was gunning for him just like the forest ranger had tried to kill him the year before. He was fighting for his life and he knew it." Hosea turned to me with a question: "If you had been the grizzly what would you have done?"

After I wrote a piece on Thayer and Sarber for the *Juneau Empire*, I received numerous emails from people who knew Sarber and his story. Some were family members, others were old-timers from Petersburg, and one man—who'd been a teenager at the time—had been part of the search party that looked for Sarber when he had gone missing. It turned out that he wasn't killed by a bear, nor did he wander into the rainforest, compelled by something bigger than himself, and allow the wilderness to swallow him.

Sarber disappeared while on patrol for the US Fish and Wildlife Service in the Bay of Pillars on the west side of Kuiu Island. New laws protecting salmon had come into effect, and Sarber was monitoring commercial fishermen in the area. When he failed to return home, a massive search party was launched. All that was found was Sarber's skiff. It was suspected that a commercial fisherman had murdered him and sunk his body.

ONE LATE APRIL, I LEFT MY HOME TO KAYAK AROUND ADMIRALTY ISLAND. My plan included exploring Eliza Harbor in the hopes of better understanding what had happened to Thayer and Sarber. I saw no boat or floatplane traffic. The loudest sounds were the occasional roaring of sea lions, breathing of whales, and squeaking of harlequin ducks as I slowly paddled south. At Point Gambier, the ocean turned rough and a humpback, seemingly upside down, frothed out of the water. A moment later, four killer whales surfaced and pushed the humpback down before it could breathe. I watched a dark bear near Eliza Harbor, its muscles rippling in its thick fur as it rooted beneath logs that winter storms had thrown up above the high tide line. The bear paused and glanced around before using one paw to shift a log and, with its other paw, reach beneath. A few hours later, I paddled up Eliza Harbor to where the bay narrows. It would still be a four-mile paddle to the head of the inlet, and a few miles more of bushwhacking to where Thayer had died. The ocean whispered and a stray gust of wind blew my kayak back. It felt like something was telling me to keep out. I studied the dark forest a few moments longer before turning south and paddling away.

CHAPTER 9

The Ghost of Hasselborg

Yet Hasselborg's knowledge of bears went far deeper than the understanding of an experienced sportsman or naturalist. He was possessed by them, from his childhood, when he had a pet racoon (a distant cousin of the bear), to just before he died, when he dreamt that bears were coming to take him away. On more than one occasion people heard him talking to bears and the bears responding as if they understood what he was saying.

—John R. Howe, *Bear Man of Admiralty Island*

In the winter of 1899, a young recluse and adventurer named Allen Hasselborg stood on a dock in Seattle trying to figure out how to get to Alaska. The Klondike Gold Rush was basically over, and while few men struck it rich, there were jobs mining, fishing, and logging in the north. Hasselborg had crisscrossed the country looking for adventure and economic opportunities, the latter of which he often found at giant mines. Alaska seemed like the next place to try his hand. While he was staring out at the Pacific Ocean, the story goes, someone knocked him unconscious. When he woke up, he was on a schooner bound for the Bering Sea to longline for cod. He'd later shrug off being shanghaied, saying he'd wanted to go to Alaska anyway.

In November of 1903—after a stint working at a mine on Baranof Island in northern Southeast Alaska—Hasselborg set out in a small dory he'd built himself to row two hundred miles from Sitka to Juneau alone. He traveled through nasty weather and rapidly diminishing daylight around the southern tip of Admiralty Island. Then he entered a giant inlet known as Seymour Canal and passed a bay named Mole Harbor. Perhaps he had a premonition, but it's just as likely he had no inkling that his life would become intertwined with the island and its bears, or that he would live alone in that bay for nearly four decades. He would become a legendary bear hunter, hermit, and key force in the movement to protect brown bears in Alaska—but right then, he was just a young man trying to stay alive in the wilderness.

In the spring of 2013, I set off from Juneau in hopes of kayaking to Sitka and visiting the site of Hasselborg's homestead in Mole Harbor along the way. It was a pilgrimage in a way. During my teenage years I'd idealized Hasselborg, even believing the myth that he never fell during all the time he spent traveling in the woods. He was basically self-sufficient, gardening, fishing, and hunting to meet the majority of his needs. He didn't bother with tents, instead relying on lean-to shelters he built or just sleeping out. He was self-educated and well versed in literature, science, and history—so much so that he left a lasting impression on renowned cultural anthropologist Frederica de Laguna after a single conversation. Hasselborg killed brown bears with an open-sight rifle—one rumored count puts the number of dead bears at three hundred. He was charged several times by wounded animals and mauled twice.

It was April 20, late enough in the year, I hoped, for the Cross Admiralty Canoe Route to be free of ice. A blizzard—an omen that would set the tone for the next ten days—set in as soon as I began crossing Stephens Passage. I traveled by compass and the angle of waves through thick, eerie fog and snow. Late in the day, the storm slowly died, revealing the snow-covered rainforest of Admiralty Island. Flocks of long-tailed, goldeneye, and white-winged scoter ducks, many recently arrived from the south, parted atop the ocean as my kayak slid past. In the dusky gloom in Oliver Inlet, a skinny bear emerged from the forest, post-holed through snow, then dug in a heap of kelp along the high tide line. Quietly, I glided over the water toward it. The bear paused, and we studied each other. It walked to the water's edge and lifted its nose to try to find my scent. I let the bear be and began portaging my kayak and gear to the northern end of Seymour Canal.

During my first night on Admiralty Island, a stiff southeasterly wind set trees creaking, moaning, and whistling. In the early dawn, I watched a small deer pause amidst quivering yellow grass poking through a blanket of wet snow. I filled my cook pot at a creek as the deer disappeared back into the forest. Later, I walked across an expansive tidal flat, pausing to sip coffee and watch a pair of shrieking greater yellowlegs sandpipers. Near the mountainside, the hooting of sooty grouse drifted over the wind. I hiked, wading through brush and deep snow, toward a giant spruce tree where a grouse was perched. His head bobbed in the brush near the top. Knowing my shotgun, with its twenty-inch barrel, would at best wound the distant bird, I passed on the shot and moved on to the next grouse. I found him in a maze of branches, silhouetted thirty yards away. Back at the beach, I boiled him with a bit of salt, made another cup of coffee, and had a late breakfast.

I waited another day and a half before the wind slackened and the ocean became manageable for my kayak. Clouds hung low; a light drizzle and my boat were all that upset the mirrorlike surface of the ocean. A small brown bear rounded the bend and trudged along the shore, hardly giving me more than a passing glance though less than thirty yards separated us. Deer, ribs protruding after the long winter, paused from eating seaweed and stared as the outgoing tide rushed me south. Curious harbor seals, with their large, dark, melancholic eyes, trailed behind me. In the afternoon as the wind and seas grew, I passed Pack Creek and made camp next to a musky mink den on Windfall Island. If the weather held, Hasselborg's homestead would be an easy, twenty-mile-day's paddle away. I stared out at the mountains rising in the gloom above Windfall Harbor and wondered how Hasselborg had contended with loneliness.

By 1903 the young wanderer had become increasingly disenchanted with civilization. He dedicated himself to trapping and living off the land, and he soon turned to hunting brown bears for their hides and for scientific collectors. A good brown bear hide was selling for fifty dollars in Juneau, equating to a month's salary working at a mine or on a fishing boat. Tired of barely scratching by and working in the brutal, dangerous conditions of giant mines, it's no wonder Hasselborg and other men turned to bear hunting. Southeast Alaska's brown bears were well culled; those that were left had retreated deep into the woods and mountains. The population was so low that supposedly Hasselborg didn't see one during his first four years in Alaska.

In 1907 Hasselborg was invited to guide the Alexander Expeditions, which consisted of a team of naturalists collecting Southeast Alaskan fauna to catalog for science. The expedition was led by Annie Montague Alexander, a quiet force who, among other achievements, founded the Museum of Vertebrate Zoology and the University of California Museum of Paleontology. Alexander worked under C. H. Merriam, who, with his zeal for brown bears, had made clear his hope that the expedition would bring back many bear specimens.

Hasselborg's aptitude for killing bears by then rivaled Ben Lilly's. Merriam's demand for brown bear specimens was so great that he was offering forty dollars a skull to anyone, and according to Hasselborg, everyone was out hunting. Hasselborg took the job, and he enjoyed the work they were asking of him, although he was skeptical of some of his employer's scientific theories. C. H. Merriam often denoted a new species of bear from variations in a single hide or skull. He claimed there were five different species of brown bears living on Admiralty Island. Hasselborg once killed two bears next to each other that he thought were brothers or, at the least, cousins. Merriam, upon receiving the two bears' hides and skulls, denoted one a grizzly and the other a brown bear. Not only did the Alexander Expedition pay Hasselborg a good salary—now, whether he liked it or not, he was well connected and would be sought out as a guide for the rest of his life.

In 1917, after more than a decade of commercial bear hunting, Hasselborg settled in Mole Harbor on Admiralty Island and began clearing a homestead. He'd remain there for nearly four decades, alone except for the hunters and photographers he guided during warmer months. He only came into Juneau, the nearest town to Mole Harbor, when he had to—which was once or twice a year. One bear guide wrote that Hasselborg often rowed a small skiff on the 150-mile roundtrip errand. In the spring, he'd bring in the furs he'd trapped during winter to sell, then buy a few provisions, pick up his mail, and stay only long enough to reply to letters. Rowing or motoring south, down Gastineau Channel, he was soon alone again with the ocean and rainforest.

In 1926, John Holzworth, a lawyer from New York, visited Hasselborg at Mole Harbor and hired the recluse to guide him on a bear hunt. In *Bear Man of Admiralty Island*, John R. Howe wrote that Hasselborg was not impressed with the New Yorker; he considered him a drunk, a lousy woodsman, and a liar. Nonetheless, he agreed to guide Holzworth again in 1928 and 1929, to film bears on Admiralty Island. Shooting bears with a camera was becoming popular, though from a guiding standpoint, it was more difficult than hunting. A cameraman had to be close to get good footage. Some early films of brown bears ended with the animal being killed by a nervous guide. Not so for Hasselborg. From his countless close calls, he'd learned to talk agitated bears down. Holzworth said his guide's words were more effective than bullets. Hasselborg inadvertently showed Holzworth that the common representation of brown bears as ferocious killers was far from the truth. Instead, Holzworth realized, the brown bear possessed a surprising capacity for reason and just wanted to be left alone.

In 1930, Holzworth published *The Wild Grizzlies of Alaska*, which centers around his time with Hasselborg on Admiralty Island. At the very beginning of the book, Holzworth wrote of Hasselborg and his home:

> The island is unfrequented but for occasional fishermen, renegade Indians, and hunter-naturalists like myself. In his choice of location Mr. Hasselborg has multiplied the natural seclusion which the island itself affords. This conforms to his purpose and his character. In his youth Mr. Hasselborg

lived widely, from Florida to Seattle. He developed an antipathy, which during years of thought, has matured in profundity, against Civilization, Bureaucrats, and Women. He achieves in his residence on Admiralty Island remoteness from the first and maximum immunity from the last two.

Hasselborg claimed he was so disgusted with the book, both because of its factual errors and the way it portrayed him, that he never read past the first chapter. Yet, even though the book's portrayal of Hasselborg, Holzworth, and grizzlies might have been stretched and, at times, inaccurate, *The Wild Grizzlies of Alaska* is still an impressive work for its time.

In 1931, Hasselborg guided Arthur Pack, founder of the American Nature Association and *Nature Magazine,* on a film shoot of Admiralty Island bears. Once again, Hasselborg awed his client with his ability to safely be near bears and communicate with them. Like Holzworth, Pack also used Hasselborg as the centerpiece in his writing about brown bears, which he frequently called "Pooh," after Winnie the Pooh, as well as other similarly childish names. Once again, Hasselborg's knowledge of bears dispelled commonly held myths that Pack had believed—and in turn, Pack shared the new perspective with a national audience.

Of course, at this same time there was the push to significantly cull, even eradicate, Admiralty's brown bears and clear-cut the island. Holzworth, a self-styled naturalist and hunter, led the charge of sportsmen and conservation groups across the nation. Though Hasselborg wanted no part in the Save-the-Bear Campaign, he inadvertently, thanks in a large part to Holzworth, became its poster boy. John Howe, in *Bear Man of Admiralty Island,* wrote, "As a genuine Alaskan homesteader who had learned to get along with bears, Hasselborg had been a powerful symbol for the conservationists. Merely by associating him with their cause, they had succeeded in drawing much national attention to Admiralty Island for the first time."

Hasselborg, to this day, is remembered by many as the bear hunter turned bear savior. However, Hasselborg was so angry at being associated with the Save-the-Bear Campaign that in 1938 he declared war on bears out of spite, according to Howe. Hasselborg, seemingly unhinged, journaled about illegally killing eleven bears from April 27 to May 25 that year. It was around this time that, after shooting and injuring a bear, he was mauled a second time—the first time had been under similar circumstances. Hasselborg's character seemed to have been molded from living alone in close proximity to brown bears for so long that, just like a bear, he wanted to be left alone and respected, and to have dominion over his territory.

MY FIRST MORNING IN WINDFALL HARBOR DAWNED GRAY AND STORMY as trees danced wildly in the wind and rain and snow splattered my tent. The ocean was a mess of whitecap waves; paddling to Mole Harbor was out of the question. I paddled over to the Pack Creek Bear Viewing Area and carried my kayak up into the snow above the high tide line. Named after Arthur Pack, and established in 1934, this watershed was Alaska's first official bear viewing area. There was no bear sign in the area so I examined the dilapidated shacks, floats, and logging boom at the outlet of the creek. They belonged to Stan Price, who had lived here from the early 1950s to 1989, and whose life story has become inseparable from the place. Two northern harrier hawks, the male smaller and almost white, glided with an enviable levity on the wind. A group of seals were hauled out on a small treeless island. In a month and a half, Pack Creek would attract dozens of bear viewers each day. Right then, it felt more than remote.

In the afternoon, after looking for wildlife in snowed-in Windfall Harbor, I paddled along the shore through sleet and whitecap waves. The weather was too bad for even the bears to be out in the open. An avalanche, sounding like the engine of a passenger plane, swept down the mountain and poured into the ocean in front of me. I took it as a sign it was time to go back to camp. I made soup and a hot drink before lying in my sleeping bag, smelling mink musk and listening to trees moan and waves crash deep into the night.

At first light the sea was silent, and I launched the kayak and continued south. Chirping land otters swam over to investigate, then made faces and snorted. Multitudes of deer, many bucks with antler nubs just beginning to bud, grazed seaweed and lay resting on gravel beaches. About seven miles from Flaw Point and the entrance to Mole Harbor, the north wind began blowing hard. A humpback whale lunge-fed along the cliffy shore I was paddling, mandating a ten-minute break pinched in sloshing waves against a kelp-adorned rock wall. On Buck Island I waited for the gale to die, but it only grew in ferocity. Hundreds of black turnstone sandpipers and harlequin ducks perched on rocks in the lee of the wind. Thirty or more harbor seals bobbed off the southern tip of the island. During the night, the croaking of migrating sandhill cranes echoed over moaning trees and crashing waves. They were on their way north to nest and feed, a journey their kind is said to have been making for more than ten million years.

In the morning the wind was still howling. I crossed a spit that connects Buck Island to Admiralty Island during low tide and followed two deer feeding along the edge of alders. I cast for Dolly Varden for a while but had no luck. Thinking I heard a grouse or two hooting over the wind, I climbed an icy ridge above Buck Lake. Fresh bear tracks wended through the snow. The second bird I found was close enough to shoot. I hiked down through a forest of massive and ancient trees. When I consulted my marine radio, I learned the weather was supposed to stay nasty for days, so I decided to walk to Mole Harbor the following morning.

IT WAS THE GREAT DEPRESSION THAT ULTIMATELY SAVED THE RAINFOREST and bears of Admiralty Island. The timber industry and the related roads that the Forest Service is in charge of building and maintaining are incredibly expensive and heavily subsidized by tax dollars. Dreams of giant pulpwood sales fell by the wayside, and Admiralty remained mostly unlogged, continuing to provide the habitat bears needed. Meanwhile, the Save-the-Bear Campaign had forced major changes in Alaska's official policy toward bears. More wardens were hired, and stricter seasons and bag limits were established.

During the last decade of his life, Hasselborg's mental and physical health slowly failed him. Howe wrote that in 1952, Hasselborg "told a friend in Juneau that bears were coming too close to his cabin and he was sure they were coming to take him away." In the summer of 1954, Hasselborg left Mole Harbor and Admiralty Island forever. In Juneau he took a taxi to the airport, riding in what he called a "hell cart" (automobile) for the first time. He flew to Washington, DC—flying was also a first—to visit family he hadn't seen for fifty years. From there, he took a train south, bought a small boat, and set out on a seven-hundred-mile journey to the island of Sanibel in Florida. He attempted to live like other retired folks but was overwhelmed by the humanity, heat, and the lack of wildlife. He returned to Southeast Alaska in the summer of 1955 and was admitted to the Sitka Pioneer Home, where he died six months later at the age of seventy-nine.

IN THE GRAY LIGHT OF DAWN, A HUMPBACK WHALE CUT THROUGH FROTH-ing waves as I trudged along a cliff. A yacht used for guided bear hunts was anchored in the relative protection of Mole Harbor. More than a dozen deer fed along the edge of the forest, sometimes beneath the orange flagging that marked a marten trap line. Canada geese, wigeons, mallards, and green-winged teal ducks, their calls sounding like laughter, rested and fed along the tidal flats. I followed an old set of bear tracks up the Mole River toward Hasselborg's homestead. Around a bend, a large "No Trespassing" sign was nailed to a tree. Half-concealed in the brush beyond was the cabin of the current owner of the homestead, complete with an assortment of blue tarps. A cracked coffee cup left out from last summer or the summer before rested on a stump. Just around the bend was where Hasselborg's cabin and shed had once stood, but all that remained were a few decaying logs. Standing on the gravel bar, under snowy unnamed mountains towering over the forest, listening to the Mole River glide by, I was overcome with the feeling I was invading the solitude of a ghost who did not want me there. I turned and followed the river back to the bay.

Instead of returning to Buck Island, I post-holed in deep snow through a forest of cedar and shore pines to Lake Alexander. The lake was frozen, which didn't come as a huge surprise after how cold the last week had been; a portage across the island to get to Sitka was out of the question. The booming of grouse echoed off the hills and ridges. In twilight, along game trails above the lake, I followed the tracks of a large bear. His prints were so fresh, I could smell him and half expected a confrontation. Kneeling, I examined the bones and hair of the remains of a deer he'd paused to smell. As I rose, I called out to let him know I was near.

The following morning rain and snow fell through the boughs of spruce and hemlock trees rising into churning mist. A red-breasted sapsucker woodpecker pecked on a dying tree as the hooting of the grouse I was seeking grew louder. I replaced the slugs in my shotgun with #2 shot when I spotted the grouse puffed up, high in the maze of branches of a spruce tree. At the echo of the shot, it plummeted into the brush. Holding the bird's warm body, I thanked it for becoming my food and watched the luster of its plumage fade. Through breaks in the trees, the frozen lake glowed white against the darkness of forest. I listened to the trees thrashing and moaning in the wind. At that moment, anonymous and deep in the forest of Admiralty Island, I was as close to knowing Hasselborg as I would ever be.

CHAPTER 10

Wilderness Lost

Those bears in the National Forests of Southeastern Alaska, especially the unique Shiras "black browns" of Admiralty Island, are being driven from one timbered retreat to another as logging crews in the employ of Japanese-owned pulp mills are permitted to bring big trees crashing down with callous disregard for other assets on this famous island, once proposed by President Teddy Roosevelt as worthy of national recreation status for all the people of America.

How are the bear faring? Not so good, especially on Admiralty Island.

—Frank Dufresne, *No Room for Bears*

Petersburg bear hunting guide Ralph Young is mostly remembered by old-timers and other guides as being rough and a drunk. One woman called him a man's man, a bear hunter. Young sometimes spent summers living with bears on Admiralty Island. He ate what bears ate. He traveled their trails. He slept in their day beds and dens. During all that time, he never fired a shot at a bear. He reveled in the peace, something that evaded him when he was in town. As an old man, he wrote that he hoped there was a bear somewhere in the rainforest with his number on it. The sudden rush of a bear, the intimacy and horror of its touch, and the possibility of being consumed: there was no way to get closer to the bear. Maybe it was a dream of reciprocity and atonement for the hundreds of bears he'd taken part in killing. Over the years, Young tried to pay his debt to bears and wilderness by fighting to save Admiralty Island from being clear-cut.

Despite the continuing change in the nation's sentiment, Alaska's bear wars didn't end in the 1930s. The idea of deliberate eradication had become increasingly unpopular, even insane sounding, but the blueprint to develop the territory was still focused on natural resource development at just about any cost. Alaska became a state in 1959 and, understanding the resource the brown bear was to the hunter, photographer, and viewer, officials were managing the animal responsibly. With stricter hunting seasons and bag limits, the brown bear's population was growing. Around 1964, in the spirit of the Wilderness Act that had just been signed into law, the Forest Service published a handbook that purported, "Visitors and onlookers from the other 49 States will be concerned with the preservation of the wildlife, scenery and wilderness for which Alaska is famous. If and when this last frontier is conquered by 'progress' something precious in the minds of the people will have died."

Meanwhile, despite the agency's flowery prose, the Forest Service was quietly working to conquer the "last frontier . . . by progress." The Forest Service's dream of turning Admiralty and much of the rest of the Tongass National Forest's ancient rainforest into pulp for paper had not changed since the death of Jack Thayer. In 1964, when Ralph Young motored to Whitewater Bay, on the southwest side of Admiralty, he was shocked by the devastation he saw. A year before, the bay had been pristine and filled with bear, deer, salmon, and other wildlife. Now, it was like a bomb had gone off. The forest had been clear-cut to the edge of the stream, destroying the spawning grounds and the watershed's salmon runs. The bears had been pushed into other watersheds, where many, particularly cubs and subadults, would fare poorly and likely be killed by larger bears.

Young motored home and, ignoring friends who claimed the loss of Admiralty Island and the brown bear were inevitable, began his career as a writer and wilderness activist. He wrote an article for *Field & Stream* entitled "Last Chance for Admiralty," which was so damning of the Forest Service's logging practices and the destruction of natural habitat that it caused the agency to issue a seven-page response in denial of his charges.

Many locals, seeing the timber industry as a way to boost the sagging economy, were mad at Young. Some wanted to fight, which Young was okay with, even going so far as to encourage the use of guns instead of fists.

Around this time, cattle owners leasing land adjacent to the Kodiak National Wildlife Refuge received authorization for the use of a Piper plane mounted with a .30 caliber M 1 Garand semiautomatic rifle to thin out bears on the northeastern section of the island. Local hunting guides contacted *Outdoor Life* in 1964, who brought the issue to the attention of a national audience. In addition to their worries about bears being gunned down from the air, guides and conservationists were concerned that cattle owners would soon be granted leases in the refuge, which would further endanger the island's bear population. After a year of operation and overwhelming public disfavor, Kodiak's aerial bear control program was shut down. In 1965 Frank Dufresne wrote the book *No Room for Bears*, which, through colorful storytelling, further raised awareness that Alaska's brown bears could easily share the same fate of their kin in the Lower 48 unless significant measures were taken.

Young spent two years touring the Lower 48, showing wildlife movies and giving lectures on the importance of conservation. Despite both small and seemingly large victories, "progress" possessed an insatiable appetite. At the end of the 1960s, Admiralty Island's future still looked bleak. Instead of a brown bear wilderness, the Forest Service and timber interests wanted roads and clear-cuts.

Ralph Young sat alone at a bar in Petersburg well before five o'clock. His wife had recently left him on grounds of incompatibility of temperament. He wrote, "I loved my wife, but I loved the bears more. That's legal grounds for divorce in Alaska and perhaps elsewhere as well." He'd become a drunk and grown fat. The wilderness he'd known in his youth was lost. The high adventure of stalking brown bears while guiding famous trophy hunters like Jack O'Connor was mostly, at best, a bittersweet memory. Then, something like a switch turned inside of Young. He left his unfinished drink on the bar, got up, and walked out. Though it was late in the year and snow was creeping down the mountains, he put together a kit and boarded his skiff. He motored out into the swift current of the Wrangell Narrows, past Sasby Island, and out into the slate gray of Frederick Sound. He dropped anchor in a protected cove of a bay that he purposely didn't name when he wrote about it later. Inside the forest's edge were the relics from a camp he had once shared with his friend and seal hunting partner, a man he referred to as "Sock-less George." George had gone missing, swallowed by the ocean or the land some years prior. Young shot a seal for food and for a number of days slept out and wandered. He looked for and found gold, then decided he didn't want it. He almost died in the crush of a maze of icebergs and thought that maybe he'd felt the hand of God guiding him through the long night.

The following year, Young's health failed and he was forced to give up guiding. He had open-heart surgery and left Alaska to live out his remaining days in Florida. Young never stopped thinking of Admiralty though. He wasn't alone in his love of the place. Tlingit elders from Angoon, other bear hunters, and conservationists from Juneau to New York fought tooth and nail to keep the island wild. In 1970, bear hunting guide Karl Lane, along with the Sitka Conservation Society and the Sierra Club, filed a suit against the Forest Service to halt the logging of Admiralty Island. The judge ruled against the plaintiffs, stating that Lane could take his clients to another island to shoot bears. Lane appealed and began a court battle that would last nearly a decade. Meanwhile, the Endangered Species Act became law in 1973, and not long after, both the wolf and the grizzly of the contiguous United States were placed on it. In 1978, President Jimmy Carter, using the Antiquities Act, declared Admiralty a national monument. A few years later, Congress voted to give much of the island monument and wilderness status and protect it from logging, mining, and road development for the time being.

In 1982, Young—who, according to doctors, was supposed to be long dead—returned to Alaska. He stepped off a plane in Petersburg and, as quickly as he could, borrowed a skiff and headed out into the wild country.

Nearly eighty years old, he motored north, relieved to be on the ocean and surrounded by rainforest. Despite the protections in place on much of Admiralty Island, Young was disturbed to see the changes to his old home. Many watersheds he'd once hunted and trapped had been clear-cut. The Forest Service had organized massive timber sales across much of Southeast Alaska and was planning on logging more. Thousand-year-old trees were being turned into pulp to accommodate Asian market demands. Communities largely made up of out-of-state loggers boomed and then busted, leaving behind ecologically and economically poor landscapes.

Young made one more journey into the wilderness, this time accompanied by a young man named Klas Stolpe. It was 1984, and Young wanted one last visit to the places that for a half century he had hunted and explored. A friend of Young's, worried about the old man being by himself, had insisted on sending Stolpe along to help. The two spent around a month and a half traveling. It wasn't an easy trip—Young was ornery, and Stolpe was new to brown bears.

"He didn't want me there, and he let me know," Stolpe, a reporter living in Sitka when I interviewed him, remembered. "He wanted to be alone, but he needed my help to do things like get out of the skiff. But when he got in the wilderness, he became a different man. He was incredibly intelligent. He could look at bear sign and tell you exactly how big the bear was and what it had been doing. He knew all the plants. He was just an incredible woodsman. It was a remembering sort of trip. He'd always be looking around, and it was like he was remembering all the hunts and time he spent in these places. Here's this frail old man, and then we're false charged by a bear and all of a sudden he turns into this great big man calmly talking to this bear. The bear walked off into the brush, and we continued up the stream."

IN MAY OF 2018, MY FRIEND MIKE JANES AND I HIKED UP A WATERSHED ON Admiralty Island I'd wanted to explore for years. Mike's wife had called me a month before and asked me to take him on an adventure in honor of him turning forty. We walked through a series of meadows and past a bear grazing near a few deer. Grandfather trails crisscrossed fields of deer lettuce, ferns, and flowers. I paused to study the edge of a bed where a bear had laid a deer skull, numerous clamshells, and salmon bones. We pushed through brush, climbed over fallen logs, and traveled bear trails deeper into the woods. During the entire day, we found no boot tracks, flagging, or cut limbs—zero sign of other people.

At dusk Mike and I pitched our tent near where the creek turned into a gorge. Rain began to fall, and we ate dinner in silence. I stared out at the gloomy forest and thought about getting older. I thought of Young and how his one true love had been brown bears—but that he had made a life from killing them. I was not sure how to navigate this contradiction. I listened to the woods until my eyelids grew heavy and I fell asleep.

In the half-light of morning, we watched a herd of deer slowly moving through the woods and grazing before we packed up camp and headed to the interior of the island. Cedar trees rose into mist, and the steady hooting of grouse filled the forest as we hiked over a low pass and down to a lake. Mike and I inflated packrafts, then paddled and portaged across a series of lakes. Mountains appeared and then disappeared in the swirling fog. That night hard rain began to fall, and the wind howled through the trees. The next morning we hiked out to the ocean and followed the beach north. A brutally muscled male bear, mostly black with a little red on his front legs, swaggered along the tide line. It was a Shiras bear, denoted by C. H. Merriam as being its own species of brown bear. The Shiras bear was used in the battle for Admiralty Island much in the same manner that the Spirit Bear—the white American black bear of the Haida Gwaii in British Columbia—has been used in the fight to protect the Great Bear Rainforest, although biologists say both bears are simply animals whose color differs more extremely from other members of their species. Nowadays very few people know what a Shiras bear is, but to me, it represents the story of the fight for Admiralty.

The bear smelled Mike and me and, in a slow and dignified manner, walked into the woods. We climbed through brush and over cliffs until we found a good beach and made camp next to rows of stacked crab pots. Seals and loons floated on the calm water of the bay. The quiet was disturbed only by the plaintive cries of mew gulls and the booming calls of sooty grouse.

In the morning we made it to Pack Creek, where a small army of tourists and their guide, who'd just been flown in from a cruise ship on floatplanes, stood in the breeze and drizzle, looking lost and confused. A young woman wearing a parka with a faux-fur hood approached and asked where we had come from and what we had been doing. I told her I wasn't sure, and she looked at me strangely. She then asked if we'd been camping out and sleeping with bears. I nodded and told her our old ladies had given us a hall pass.

Mike and I spent a few hours in the observatory, enjoying a close walk-by from a big male bear, before catching a flight on an empty floatplane back to Juneau. I studied mountains wrapped in clouds and ocean and imagined an old, enfeebled Ralph Young walking up a salmon stream. I imagined a bear huffing and popping its jaw before charging and Young forgetting his demons and sins. I imagined him speaking to the bear and, for a short while at least, the bear gifting the old man the return of his lost wilderness.

INTERLUDE

In 1968 Doug Peacock, a Green Beret medic who had just finished serving two tours in the highlands of Vietnam, flew home to Michigan. He would later become a cult hero and an author who would write the book *Grizzly Years: In Search of the American Wilderness*. In that book he recounts how during the war he had carried a map of Montana and Wyoming. He was surrounded by madness, violence, and death, but he'd study the blank places on the worn paper and imagine mountains, rivers, and woods.

In Michigan, suffering PTSD and unable to talk to anyone, he bought a Jeep and drove west. He spent months roaming the Southwest and the Wind River Range in Wyoming before heading north to Yellowstone. He hid his Jeep off a dirt road and hiked through a lodgepole pine forest to the edge of a river. An attack of malaria came on as he made camp. Soon he was lying in his tent with a dangerous fever, wracked by chills. After a few days of delirium, the fever broke and he was able to crawl out of his tent. The following morning, while he was soaking in a warm spring, a mother grizzly and two cubs appeared. When the bears ambled closer, Peacock rose naked from the creek and sprinted for a tree, colliding with the trunk so hard that he gashed his forehead and nearly knocked himself unconscious. He clawed his way up and sat on a branch. The bears ignored him as they ate grass and dug roots. Eventually, the family wandered off and he made his way back to camp, exhausted and on the verge of hypothermia.

"Now that I knew real grizzlies lurked in the shadows, my dreams were not so important," Peacock wrote. "Something big was out there. For the first time since returning to the world, my thoughts chose themselves without Vietnam intruding."

One hundred sixty-three years had passed since Meriwether Lewis saw, then promptly shot, his first grizzly. The last vestige of that population, probably numbering in the hundreds, existed in Yellowstone National Park, Glacier National Park, and a few other isolated pockets near the Canadian border. The future of these remaining grizzlies, Peacock soon learned, was very much in peril. After encountering the mother and her cubs, he felt compelled to go deeper into the wild. On his first intentional field day observing grizzlies, he watched a procession of fourteen marching along the same bear trail to feed at a garbage dump. People were only beginning to understand how bad it is for bears to become habituated to eating human food and rubbish—it turns them into addicts and makes them associate humans with food, increasing their likelihood of messing with, and even attacking, a person. The August prior, in Glacier National Park, two nineteen-year-old women, camped miles apart, had been killed by two different human-food-conditioned grizzlies during the same night. No bear-related fatalities had been reported in the park's fifty-seven-year history before that. It was a strange coincidence that still leaves people wondering, but regardless, both deaths would have been avoided if the bears had not learned to associate people with food.

Peacock spent the next month in Yellowstone. One of his most profound experiences occurred when he unexpectedly blundered near a dominant male bear in the area. The bear gnashed his teeth and lowered his ears, indicating he was considering attacking. Peacock aimed his pistol at the bear's massive head, but instead of pulling the trigger, lowered the gun. After several moments, the bear calmed and looked away. Peacock did the same, then took a step back. Next, he wrote:

The grizzly slowly turned away from me with grace and dignity and swung into the timber at the end of the meadow. I caught myself breathing heavily again, the flush of blood hot on my face. I felt my life had been touched by enormous power and mystery.

I did not know that the force of that encounter would shape my life for decades to come. Tracking griz would become full-time work for six months of many years, and it lingers yet at the heart of any annual story I tell of my life. I have never questioned the route this journey took: it seems a single trip, the sole option, driven by that same potency that drew me into grizzly country in the beginning.

At the time, Peacock likely didn't understand the significance of his not shooting that bear. He just knew he found it easier to talk to bears than to priests. In the narrative of America's relationship with brown bears, that moment represents a paradigm shift, a new perspective for future generations to consider. Grizzlies and the wilderness saved the veteran's life, and in return he would spend the rest of his life, a half century and counting, learning from them and advocating on their behalf. Meriwether Lewis's ghost still haunts grizzly country, but Peacock acts as a path of hope for the future of people, bears, and wild places.

Part II

CHAPTER 11

Pray

It is true also that certain experiences, states of mind, and ways of life, cannot be willed back. That intuitive relation to the world we shared with animals, with everything that exists, once outgrown, rarely returns in all its convincing power. Observation, studies in the field, no matter how acute and exhaustive, cannot replace it, for the experience cannot be reduced to abstractions, formulas, and explanations. It is rank, it smells of blood and killed meat, is compounded of fear, of danger and delight in unequal measure. To the extent that it can even be called "experience" and not by some other, forgotten name, it requires a surrender few of us now are willing to make.

—John Haines, *The Stars, the Snow, the Fire*

I've been fascinated by brown bears for as long as I can remember, but it wasn't until 2010 that I really began questioning the nature of our relationship with the animal. At the end of July, MC and I drove to the Arctic, where she dropped me off to begin a seven-hundred-mile journey across the Brooks Range. She then drove south to enroll in a creative writing program at the University of Nevada, Las Vegas. We hadn't known each other long, and I figured that was the last I'd see of her.

Not even an hour into my hike, I watched a reddish grizzly emerge from a willow thicket swinging its head back and forth as it walked down the creek bed in my direction. I crouched, unnoticed, and the bear ambled across the creek, climbed up a small canyon, and lay down under a stunted black spruce tree. I hiked on until I startled a wolf standing at the edge of a willow thicket. It leapt away, then froze and glanced back. One of its paws was mangled. Its head looked huge in comparison to its emaciated body. Slowly, on three legs, its ribs jutting out of its mangy gray coat, the wolf approached. I yelled, and it hopped back into the brush.

The wolf followed me, whining, barking, and letting out ominous howls. A few minutes passed before it emerged from the brush and cut me off. It glanced around nervously then walked toward me. I yelled and waved my arms over my head. It stopped for a moment and then came closer. I had brought my brother Luke's .357 pistol—it was light for bears but, at the time, was the only handgun I had access to. I pulled the pistol from the holster and yelled as angrily as I could. The wolf froze, glancing skittishly back and forth, before hopping into the willows. I continued hiking, and the wolf paralleled me. Twenty minutes later, it cut me off and approached again. At thirty feet, I leveled the pistol and yelled. It glanced back and forth, looking both desperate and terrified, then hobbled closer. At twenty feet, I aimed the pistol.

"Leave me alone!" I yelled as loud as I could.

I'm not sure why I didn't pull the trigger. Maybe it was because I was confident it could not hurt me, armed as I was. Maybe it was because I don't like killing animals I don't eat. Maybe it was because I tricked myself into thinking there was hope the wolf would somehow survive. Maybe it was because I lacked the compassion to help it die.

"You should go now," I told the wolf.

The wolf hesitated, glanced around, and then slowly hopped on three legs into the willows. It climbed above the creek bank and followed me for the next three and a half hours. Its howling kept me on edge until it wandered up a valley and left me to silence. I'd planned to travel the same valley the wolf was howling from, but I rerouted and hiked over two mountains instead.

At two in the morning, I stopped and studied mountains and valleys bathed in soft light.

"Okay to camp here?" I asked.

I'd dreamed of making this trip since 2004 when I first wandered into the Brooks Range, but now I was questioning my plan. I'd had close calls with bears earlier in the summer on the ABC Islands—one mom with tiny cubs had come very close to charging after I stumbled into her. I didn't know how to talk to bears then, but intuitively, I had sat down and waited until she calmed. Now, a small blond grizzly feeding on blueberries appeared far below. I wasn't sure, but I thought the bear was in the same valley where I'd been charged, and then spared, by the grizzly during that first trip. I watched the blond bear for a while and pushed away all thoughts of quitting the trip.

When I finally fell asleep, I dreamed of starving and being eaten by a progression of different animals. In one chaotic sequence, an emaciated, absurdly well-endowed bull moose emerged from a Technicolor forest and approached. I was armed with a popgun and was unable to move as the moose's hulking body covered me in darkness. I woke frequently and could not shake the feeling there was something outside the tent. I glanced out the vestibule but saw nothing in the low light.

"Thank you for not killing me," I said to the world as I broke camp.

A wolverine streaked across the mountain to a clump of willows—I walked within feet of it before it noticed me and ran. Herds of Dall sheep appeared on the surrounding mountains. I became aware of a droning throbbing in the heavy silence that grew stronger the deeper I traveled into the country. Near the base of Mount Doonerak, I encountered a grizzly, followed by two cubs that were nearly her size, lumbering around the corner of a ravine. They were thirty yards away but looking in the opposite direction. I backtracked, then climbed out of the gully up onto the scree slope of the mountain. The three bears climbed the opposite side of the narrow canyon and then stopped and stared at me. The mother raised her hackles, the hair on her hump stood up, and she let out a series of deep woofs. I did not know what else to do, so I spoke to her gently, saying I was sorry, that I would not hurt her or her cubs, and asking her to let me pass. It felt weird, but after a while she stopped posturing and huffing, and followed her cubs up the mountain.

On my fifth day of travel, a small, ragged Nunamiut man limped down Ernie Creek toward me. Horror, shock, and then relief flashed across his face. His soiled blue jeans were torn, and his sweat-encrusted, holey cotton T-shirt offered little protection against the cold. For a moment I was embarrassed; I had to ford creeks and rivers so frequently, I'd given up on wearing pants.

"Thank God!" he cried. "I'm happy to see you! Am I headed toward Anaktuvuk Pass?"

"No," I said, checking my map, suddenly worried I was going the wrong direction. "You're heading toward Coldfoot."

The man could not remember how many days he'd been lost or how long ago he'd stepped out of a car at Coldfoot, a truck stop on the Haul Road, and begun hiking northwest. The sole of one of his shoes had fallen off. He'd been out of food for days and had lost most of his clothes. He did not recognize the wide treeless valleys and rugged mountains around him, even though they had been his people's home for hundreds, if not thousands, of years. It had been nearly two decades since he'd left Anaktuvuk Pass, nestled between mountains where the John and Anaktuvuk Rivers are born, to live in Fairbanks.

I passed him my hat, an extra fleece jacket, and a pile of candy bars as he muttered, "Thank God," over and over. He devoured the candy, throwing wrappers in the creek and in the bushes, then asked how to get to Anaktuvuk Pass.

"It's that way, maybe forty miles or so," I said, pointing the opposite direction from where he'd been traveling. "You can walk with me."

He thanked God again and we continued up the drainage. At our second creek crossing, I suggested he also leave his pants off to save time. He looked at me strangely, then laughed. Soon we were walking up a canyon in our underwear, toward a lonesome sweep of gray and black mountains.

In the late evening, after climbing a steep pass and crossing the Continental Divide, we sat near the headwaters of Graylime Creek, close to a small fire, looking over a dreamlike expanse of mountains, valleys, and sky.

"I don't know what would have happened if you hadn't found me," he said. "Guess I just would have walked back to Fairbanks."

The fire burned brighter as he heaped another armload of dry willows atop the flames. He pulled out a Nalgene bottle stuffed with weed.

"Can I see your map?" he asked.

He tore off a rectangle and rolled a generous joint.

"Wish I had a bottle. Drank mine the first night out. Best thing about living in Fairbanks is booze is cheap. I get a drink, and I just feel better about things," he said.

We shared a dinner of candy bars, crackers, and Tang as we talked about the country, its animals, and people. Although he hadn't encountered any bears, he was terrified of them, so much so that a can of bear spray, with the safety off, was always in his hands. Before we turned in for the night, he said, "I hope they're looking for me. Tired of walking. Too far. Too many bears."

The people from Anaktuvuk Pass, the furthest inland Iñupiaq village in Alaska, call themselves Nunamiut. Before Euro-Americans came to Alaska's Arctic, the Nunamiut population was probably around a thousand. Near the end of the 1800s—with the arrival of commercial whaling ships, explorers, and missionaries—famine, disease, and alcohol devastated them. New trade items altered bartering and subsistence patterns. At the turn of the twentieth century, the caribou disappeared and the Nunamiut starved, their population shrinking to around one hundred. The survivors were forced to join their Iñupiaq cousins on the coast or migrate hundreds of miles into Canada, where caribou populations remained healthy. In the late 1930s, when the fur industry collapsed and the caribou population began to recover, a handful of disenchanted Nunamiut families left the coast and returned to the central Brooks Range. They lived much as they had for hundreds of years: moving across the country, harvesting with the seasons, mostly cut off from and virtually unknown to the rest of the world. They built deadfall traps to kill small mammals, snared ptarmigan, sewed caribou skin tents, wove fishnets, killed wolves for bounty, and as they had for centuries, hunted caribou. Years elapsed, and Nunamiut families began to settle into the mountain pass separating the Anaktuvuk River from the John River. A post office was established in 1951, followed shortly by a trading post, a Presbyterian church in 1958, and a permanent school in 1961.

The following afternoon, two Nunamiut men on four-wheelers drove up the wide tussock-covered valley. They hugged my companion, loaded a pipe with a big bud, and took a few tokes to celebrate his safety. Using Nunamiut words, they spoke of grizzly bears, caribou, Dall sheep, and wolves. Both had AR-15 assault rifles strapped to their backs.

"I kill anything up to two hundred yards," the younger man told me.

"I'll piss anything off under ten feet," I said as we exchanged guns. He looked down the pistol's sights and grunted with empathy.

"You want some caribou?" he asked, pulling out a plastic grocery bag full of boiled meat.

My companion climbed onto the back of a four-wheeler and was about to drive away when I asked for my hat and jacket back. After our goodbye, I sat on a tussock in a humming cloud of mosquitoes, eating ravenously. Loneliness came but passed a moment later when I felt the caribou's energy. The drone of the four-wheelers faded into silence as the three men dissolved into the windswept tundra. Jagged mountains rose into darkening clouds. Slowly, I became aware of the pulsations reverberating in my chest and ears. It was as if the land possessed a giant beating heart that could be heard only in solitude and stillness.

A storm rolled in, graying out the mountains; rain began falling, and wind began to howl. I usually make a point of camping in the open, since bears often travel in the brush and are emboldened when they have cover. But at midnight, soaked and chilled, I made an exception and set up my tent in a willow thicket.

A few hours later I awoke yelling, "Hey! Hey! Hey!"

Without knowing what was happening, I tore open the tent's vestibule and cocked my pistol as a grizzly let out a huge whooshing grunt. The bear just missed me as it rushed over the tent and crashed off into the willows. I stood in the gloom waiting to see if the bear would return. Willows danced wildly in the wind, and rain pattered and pooled on the collapsed tent. After ten minutes or so, I spoke. "Thank you for not killing me. I'm leaving now."

THE FIRST MAN I SAW IN ANAKTUVUK PASS STOPPED HIS FOUR-WHEELER and, after a short conversation, gave me some dried caribou meat. The village of three hundred people consisted of a precise grid of trailer-like homes, most complete with satellite dishes and caribou antlers. Stopping at the post office, I picked up a box of food I had mailed and then ran into the man I'd been walking with. He was wearing the same clothes he'd been hiking in and still seemed lost.

"Come visit us," he kept offering.

I said I would after I went to the store, but knowing he had brought more than just one bottle of weed into Anaktuvuk Pass, I snuck out of the village instead. A few miles later, I came across a pile of caribou skulls and waist-high antlers lying in a pile. A piece of weathered steel protruded from the surface nearby. Grabbing hold, I pulled an ancient wolf trap from the lichens. Testing its springs, I doubted its capacity to hold a wolf anymore. I wondered if it was left over from wolf bounty days, when the Nunamiut could make a fair income hunting and trapping wolves. A few hours later the clouds tore open and rays of light glistened on the wet tundra. In twilight, I made camp above the John River.

The following day, I encountered numerous caribou hunters from the village out on four-wheelers and Argos—an eight-wheeled version of a four-wheeler.

"Do you have any boots? Where are you going?" one elder asked as he studied my sandals.

"I'm hoping to walk to the Noatak River. A couple of my friends are going to fly in and meet me there with a canoe, and we're going to float out to Kotzebue," I said.

"You have a long journey before you. How come your friends aren't with you?" he asked.

"They're smarter than me," I said.

A young hunter talked about the guy I'd found and how he'd brought drugs into the village. The old man winced, and a pained expression came over his face. After a few moments, he pulled out a plastic bag that must have held ten pounds of bloody meat and offered it to me. I thanked him but told him I couldn't because of bears.

"At least take this," he said, handing me a plastic bag with half a boiled pelvis inside.

I hiked up Ekopuk Creek as herds of caribou moved restlessly across the mountainsides, their clacking tendons and hooves and snorts echoing in the quiet. I noticed a circle of stones large enough to hold down the edges of a caribou skin tent on a small gravel rise—the sort of home the Nunamiut used for thousands of years. Caribou antlers and bones lay nearby. Just outside the circle, I sat and gnawed meat from the pelvis. I put a hand on one of the stones and listened to the creek rippling and the pulsations reverberating in the stillness. I tossed the pelvis next to old caribou bones and continued up the Ekopuk Valley.

The pulsations grew stronger as I walked up the valley. A cow moose, looking gargantuan in comparison to the hundreds of caribou in the area, browsed a patch of willows. A blond bear trailed by two cubs of the year appeared. I sat down, and the bear family, though unaware of me, passed so close that I could see terror in their eyes. A few minutes later, a big, angry-looking male bear rose on his hind legs from the willows. I froze. He inhaled and scanned the area before fixating on me. Several seconds passed while he tried to figure out what I was before he dropped to all fours and was again hidden by brush. I pulled out my pistol and spray, and then began walking toward the other side of the valley. I jumped small groups of caribou, tendons clacking and antlers held back as they fled through the brush. A few minutes later, the bear rose to his hind legs again, this time looking even angrier. I crouched and froze. When he dropped to all fours, I went into thick brush to get farther away. Fifteen minutes later, the willows exploded as a brown animal came my way. When I let out my breath, I realized it was two caribou bulls walking toward me, gingerly nibbling bushes, seemingly oblivious to me standing just feet away.

Caribou tracks, bones, and carcasses littered the bottom of the drainage. Bear and wolf tracks weaved from skeleton to skeleton. Their hair-filled, bone-speckled scat peppered the tundra and gravel. The creek was easier to travel along, but it was also used by bears to ambush caribou. I was tired, so I gave in, yelling constantly as I trudged along the gravel. A cow caribou and her calf walked into the creek and came closer, pausing ten feet away in a still pool that glowed in the light of the setting sun. *This is a corridor of death*, I wanted to tell her, *take your baby higher*. I clambered up to open ground and sat watching caribou move across the dimming land and listening to the pulsations. I considered different explanations for what I was hearing and, at times, feeling in my chest. It was not the beating of my heart. I took my pulse to make sure I wasn't having heart issues. Maybe the pulsations had something to do with the permafrost melting. Or oil companies developing the National Petroleum Reserve to the north. Or some sort of military weapons testing. Or auditory and sensory hallucinations I was suffering from being alone. Or maybe they were caused by something else.

Caribou filled the valleys as I hiked toward the headwaters of Ekopuk Creek. I came upon the fresh scat and tracks of a large grizzly and became inexplicably anxious. There was grizzly sign everywhere, but there was something about this particular bear that made me nervous. I hiked through a mountain pass as small herds of caribou swirled around me. I encountered more scat from what looked to be the same bear, and my nerves became increasingly shot. Edging along a small lake pressed against a mountain, I watched a few caribou moving across slopes and down ridges. I looked directly above and saw a large grizzly three hundred yards away charging toward me. By the time I pulled out my pistol and bear spray, the bear had closed another hundred yards. There didn't seem any force on Earth capable of stopping a bear its size, moving so fast. At less than a hundred yards, the terrain flattened and the grizzly crashed into a narrow band of willows. The swath of small trees wasn't more than ten feet wide, so it shouldn't have stalled the charge, but the bear never emerged.

A few hours later I set up my tent on a small knoll covered in caribou poop. Mountains to the west glittered in the soft light of the setting sun. The pulsations reverberated inside my chest as I watched a Dall sheep ewe traverse the scree-covered slope above.

On Agiak Creek, three grizzlies traveled in front of me as I hiked down the valley. They all appeared independent of each other. The bear highest on the mountain rummaged through the willows along a small ravine, while the bear partway up the mountain was darker colored and, judging by its aggressive body language, was trouble. Caribou sprinted away to give that bear a wide berth as it swaggered across the tundra. The bear closest to me was smaller and was meandering along the creek's gravel bar in the direction I was going. I edged within one hundred yards of the little blond bear and snuck by undetected.

I didn't sleep much anymore, and when I did, it was brief and unsatisfying. I woke to the slightest changes in sound and at times exploded out of my sleeping bag, pistol cocked, when a rogue breath of wind hit the tent. It was a rare day I did not see a bear, and sometimes I saw as many as five. The bears became my world. Everything else, the caribou and other wildlife, the river crossings and the pulsations, thoughts of loved ones, became secondary.

Above the muddy, slow waters of Agiak Creek, a caribou skull and antlers pointed skyward like some sort of religious icon. I watched a grizzly grazing within thirty yards of a half dozen caribou. Neither seemed to care the other was there. A little later, I sat down when a small blond bear and her two tiny cubs came running across the tundra in my direction. She stopped a short distance away and, unaware of me, began feasting on a profusion of blueberries. One cub busied itself eating as its sibling tried to entice it to play. After twenty minutes of grazing, she began her loping run again and led her cubs away across the tundra.

Thousands of caribou filed through the valley above April Creek. Often groups or even small herds going in the same direction would parallel me. Once, when I pushed through some brush, I stumbled upon the carcass of a mostly eaten caribou decomposing in a pool. The bushes exploded, and an old cow stopped and stared at me. It took a second before I realized I was pointing my pistol at her. A young bull walked with me for an hour or more up a willowy drainage. He periodically stopped and looked back as if he was waiting for me to catch up.

My sandal straps broke, and I rigged them with cord, turning them into a burly set of flip-flops. I was constantly immersed in water, sometimes above my waist, so wearing shoes was out of the question. White and red spots appeared on my swollen feet. I dodged bears, usually able to avoid them before they saw me, except on a mountain pass, where a blond bear ran toward me. It stopped at seventy yards, then paralleled my movements for several minutes. It didn't exhibit any signs of agitation but rather a sort of predatory curiosity. A few days later, I limped down the Alatna River valley as a large blond male bear with dark legs swaggered in my direction. He became distracted when he blundered into a small group of caribou and his walk became even more exaggerated. I ducked out of view and gave him a wide berth.

I camped at the pass that led to the Noatak River and had nothing for dinner, saving my last food—a Clif Bar—for the next day. I was hungry and weak, but also relieved. There were only a dozen or so miles left to the lake where I'd meet my friends, who were scheduled to fly in in a couple days. That meant only a few days with no food. In the morning the fog was too thick to travel without the risk of blundering into a bear. Around noon, I heard deep breathing and heavy steps approaching. Quietly, I slipped out of the tiny tent and stared into the swirling gray. The thing sensed me, froze, and quieted its breathing.

"I'm here," I said softly.

The creature exhaled deeply and slowly approached. A shape gradually began to appear in the murk. Soon a huge set of antlers manifested and three big bull caribou appeared. They paused for a few moments before continuing through the fog. Rain fell steadily, loosening truck-sized boulders from the mountains and swelling the creeks. The wind began to howl, tearing open the fog and revealing the rocky slopes of serrated peaks. I packed up camp, walked around a narrow lake, and hiked down into a canyon. A small grizzly appeared, grazing on blueberries. The rain matted its fur, making it look like a person wearing a bear suit. There was no cover, and we were heading in the same direction, so I traveled alongside, watching. It was so fixated on berries, it never noticed me. The remains of a large Dall sheep ram protruded from the earth. When I grabbed a horn, I was surprised by how spongy it had become. I apologized and left it alone. Numerous bands of caribou fled as I hiked along the slope. The wind and rain increased, and I became more on edge. At any moment I expected a bear or something worse to explode from the brush and come for me. I climbed out onto a low ridge and looked down toward the wide expanse of the Noatak River valley. I'd been traveling for three weeks, and I'd dreamed of making it to the Noatak through good and bad times for six years.

"I'm here," I said.

The next thing I knew, I was doubled over, unable to breathe, and sobbing. A moment later, a haunting swell crept across the tundra. For a second I was confused. Then it was clear. Wolves, four gray and one white, sat howling on a knoll nearby. I clambered back to my feet, wiped the tears from my face, and began walking toward them. They were quiet for a few moments, and then the big white wolf rose from its haunches and let loose. They were probably warning me to stay away, but in the moment, I believed they were welcoming me.

I CAN'T REMEMBER HOW MANY DAYS I SAT NEXT TO THAT SMALL LAKE waiting for my friends to show. I wondered if they'd forgotten me, or if something serious on a national or global scale had happened. At the head end of the valley, Mount Igikpak, the tallest peak in the western Brooks Range, peeked out from the clouds. A dark grizzly traveling with a blond grizzly, both looking like adults, passed across the mountain slope above camp. The pulsations grew louder and reverberated more deeply until at times they were nearly overwhelming. I passed days picking the tart, less-than-abundant crop of lowbush blueberries and trying, unsuccessfully, to figure out how to catch a pike without some line.

One afternoon I was on my hands and knees picking berries when I was startled by a rushing sound. I rolled over and yanked out my pistol, figuring it was a bear. Ten yards away, six exhausted bull caribou with their tongues lolling sprinted by without so much as looking at me. I followed the smallest with the pistol sights just behind the shoulder. I thought about ending my hunger, but I didn't want to waste meat, nor did I want to smell like blood and attract bears. A howl rose out of the tundra and was answered moments later by other wolves across the valley. They howled back and forth excitedly until they were together, no doubt feasting on a caribou that minutes before had been traveling with the bulls that had just run past me. That evening, fantasizing about globs of red meat, I limped across tussocks looking for the kill. I was exhausted and moved slowly, stopping constantly to listen for any sign—birds, the crackling of brush, the alarmed whine or bark of a wolf—that might mean I'd find a way to satiate my hunger. An hour elapsed. I heard nothing other than the pulsations rippling through the silence and reverberating heavily in my chest. I gave up and sat, despairing, on a tussock. A nearby howl roused me, and I stood, swaying as I tried to find my balance. The white wolf paced back and forth on a knoll, watching me.

My friends Ed and Ben flew in the following day—they'd been stuck in Bettles on weather hold for several days. Soon, we were floating down the Noatak River toward the Chukchi Sea. With companions and food, life became so easy that in the evening I often forgot to ask permission to camp and, in the morning, give thanks for being alive. My friends could not hear the pulsations. They made fun of me for becoming "bushy." The trip was a joy but lacked the raw magic I'd encountered while traveling alone. Every bear we saw ran. The pulsations, blotted out by the river and conversation, became nearly silent. I hiked away from camp across several hundred yards of tussocks until I could no longer hear the river. It was September now. The willows were yellow, and there was a sharpness in the air that promised snow. Slowly, I became aware of the pulsations again. They grew in pitch until they reverberated in my chest and echoed in my ears.

On our second-to-last night out, we sat late around a campfire silently watching flames flicker. After six weeks in the Arctic, I had no idea what the nature of the world was that I would be going back to. That night, unbeknownst to us, a family of grizzlies visited our camp—a mom with two big cubs. In the morning we found where they had walked maybe forty yards from where we'd been sleeping. Their tracks showed their anxiety and intrigue, how they had stopped and studied our tents.

"Thank you for not killing me," I said one last time, before we packed up camp and headed home.

THAT WINTER I FOUND MYSELF IN A WORLD WHERE BROWN BEARS MEANT nothing. MC had not forgotten me or found someone better looking and funnier, despite being surrounded by two million people in Las Vegas. I traveled to places I never thought I would visit. In Washington, DC, for the first time, I trailed MC as we toured the Smithsonian National Museum of Natural History. The place was filled with tourists, stuffed animals, bones, and models of ancient people. There was a life-sized rendition of Neanderthals burying a dead family member wrapped in the skin of a brown bear and surrounded by bear bones. On one wall were nearly fifty hominid skulls depicting our evolution going back seven million years.

The following morning, while we sat in a Starbucks above the frozen edge of the Potomac River, MC typed on her laptop and I thought about those skulls, their hollow gawks. I looked at the coffee shop's other patrons and wondered about their lives. I thought about the bears, asleep in their dens, the world above them covered in snow and darkness. We left DC and began driving to the city Hunter S. Thompson had deemed "the heart of the American dream," the city MC called "the wet dream of America," the manmade feature most visible from space, the city that represented the antithesis of who I'd come to know myself to be: Las Vegas.

We followed multilane interstates going seventy-five miles per hour, surrounded by traffic, too fast to see much of the world we were traveling through. The winter before, I'd ridden my bicycle across the country on a similar route. Even though we were traveling the same country, I was experiencing a different world. We paused in New Orleans and watched huge industrial barges, cutters, and even a giant cruise ship motoring through the muddy waters of the Mississippi River. We wandered the French Quarter listening to brass bands and drinking overpriced watery drinks. We drove on, past bayous and cypress swamps into the brushlands of Texas. We subsisted on breakfast sandwiches, becoming sluggish and constipated. We traversed the desert without consequences, listening to MC's iPod and staring out at the blur of the desiccated landscape.

We entered Las Vegas, passing beneath giant billboards advertising infidelity, blue aliens, and lawyers curious if you had been in an accident.

MC took me to the Strip, and beneath skyscraper casinos we dodged people trying to live out prefabricated fantasies. Malnourished-looking Hispanic people slapped cards together with pictures of nude women on them advertising "Hot Girls for Cheap!!!" In the casinos, thousands of people stared vacantly at slot machines. Occasionally, someone hooted and hollered as coins avalanched out. Huge glittery images of sexy people, including a woman's naked butt, and fast food flashed on giant electric billboards. Street corners were strewn with newspaper boxes full of prostitution classifieds. A guy standing in a darkened doorway invited us to come in and have free sex. On a skybridge, a one-handed panhandler played the guitar and sang Johnny Cash. Nearby, two young children clung to their mother, ragged and with dark, sunken eyes, who held a sign that read, "Please God Anything Will Help." In the shadows of the neon lights were other homeless people—deranged, desperate, and often whacked out on chemicals. On the street below, a man held a sign reading, "$20! Kick Me in the Balls! I'm Not Wearing a Cup!"

We left the Strip and drove to the house where MC was renting a room. During the night, I listened to sirens wailing down Flamingo Road and thought about desperation and hunger. I thought of the hominid skulls mounted to the wall of the Smithsonian as I and multitudes of other tourists filtered past. I closed my eyes and tried to imagine caribou moving through a valley, a wolf howling on the tundra, a grizzly charging across a mountain. I tried to recall the pulsations, what they sounded and felt like, and pressed my fingers against MC's pulse. The beating of her heart echoed into my flesh, and for a few moments, I remembered.

CHAPTER 12

The Woman Who Married a Bear

Philosopher Ludwig Wittgenstein said, "If a lion could talk, we could not understand him." Lions' language and actions, like bears', evolved from the way in which they adapted to and survived in their daily environment. Since we people face different challenges in life than do bears, it is difficult for us to understand the world of a bear. What we can do is understand the bear's world in human terms.

—Stephen Herrero, *Bear Attacks*

The winter of 2011, while surrounded by the excesses of Las Vegas, I couldn't stop thinking about bears. The following spring, I set out in a kayak and paddled 250 miles around Admiralty Island. Not long after, I began guiding bear viewing trips, which had become immensely popular in Alaska and British Columbia. Some areas like Brooks Falls in Katmai National Park & Preserve attract hundreds of viewers a day during peak season. Other areas like McNeil River, and to a lesser extent Pack Creek, are managed for more of a wilderness experience with a smaller number of visitors allowed each day. Bear behavior varies in every region, but bears that frequent viewing areas are generally used to people, and as long as visitors behave in a respectful and predictable manner, most do not appear to significantly alter their behavior. These bears can pass within yards of visitors—on occasion a subadult or mom with cubs will even seek out a person (usually someone they are familiar with) to use as a buffer against other bears that scare them.

During the summer of 2012, on a trip to an area where bears were less used to people, my clients and I encountered a woman camped alone near the mouth of Middle Creek in Windfall Harbor on Admiralty Island. My clients and I were in kayaks; I paddled closer, but the woman stepped over an electric fence and disappeared into her tent. I waited a few moments, hopeful she would emerge, come down to the beach, and talk. Pink salmon struggled upstream as Bonaparte's and mew gulls screeched and pecked at carcasses. When a floatplane picked my clients and me up that evening, I stared down through sheets of rain at her tiny tent surrounded by wilderness. The Pack Creek Bear Viewing Area, monitored by rangers, was only a few miles away. It might sound like help was close, but if a bear decided to hurt her, she may as well have been on the moon. That night at home I imagined the woman lying in her tent listening to the rain, wind, and bears prowling the darkness.

A few days later, I was back in Windfall Harbor when I encountered the woman paddling her packraft in a steady drizzle toward Windfall Creek. Unable to escape me and my clients in our much faster kayaks, she hung her head in defeat when I called out an overly excited greeting. I asked her a few questions while she stared at me with a look of exasperation. She said she was a professor from a college on the East Coast. Her purpose for being on Admiralty Island, as I understood her explanation, was to investigate the world of bears using a variety of forms of artistic methods and technologies, while trying to answer philosophical questions like whether humans have a place in the world of brown bears. She voiced her misgivings about the behavior of the few she'd seen. They came at night. She could hear their claws on the gravel, their grunts, the deep exhalations of breath, the sharp pop of salmon skulls being broken open, and the chomping of their jaws. During the day, they stayed inside the forest just out of sight. She looked out on the expansive tide flat of Windfall Creek.

"No bears in the estuary. They're all napping. You won't see any," she said and offered us a little more advice about bears. She turned her pack-raft to begin the mile paddle back to her tent. I understood then that she had no idea what she was doing. I wondered if I should mention that her camp was on the bears' dinner table and how there was only a short stretch of Middle Creek where salmon could spawn—that she was stressing and displacing bears and putting herself at risk of being mauled. A few years later I wouldn't have hesitated setting her straight.

"I'll be back in a few days. Do you want anything from town?" I said instead.

"A book of good poetry," she said after some thought.

She paddled away as one of my clients shook his head and muttered, "Crazy." We paddled to shore and hauled our boats well above the high tide line. I positioned my group at a log near where the creek entered three-foot-high beach grass and asked them to sit. I used the same spot every time I came to this creek to isolate our smell. It was out in the open to avoid surprise encounters and far enough away from where bears were fishing to not disturb the majority of them. I do everything within reason not to alter bear behavior. Marching up salmon streams leads to encounters with scared and angry bears, and can frighten animals away from a watershed for a day, weeks, even for an entire season, as can talking loudly or moving around too much.

Twenty minutes later, a young bear emerged and began splashing up and down the creek chasing chum salmon. A woman squealed in delight when the bear finally caught a fish in full spawning colors. At the peak of a strong run—in Southeast Alaska the run generally lasts from around mid-July to mid-September—bears tend to eat just the roe and skin from female salmon and the brains and skin from males. But when the bears are hungry, and during hyperphagia, they frequently eat everything but the gill-plates. One study calculated that an adult male on Kodiak Island eats more than 6,000 pounds of salmon annually. An adult male on Admiralty likely consumes less, in large part because Admiralty tends to have smaller salmon runs.

Soon, a Shiras bear swaggered out of the grass to the bank. It was larger, and its belly hung low as it prowled the creek studying the current for a fish it wanted. It walked around a bend and disappeared into the old-growth forest. Before we left, I slowly led my small group a short distance to the edge of a field of tall grass. I cautioned everyone to stay silent and close together. Bear trails, scat, and salmon remains were scattered everywhere. The sea of grass looked still. A moment later, the grass twitched. Through binoculars we made out a paw, and then a bear lying asleep on its back. Someone said something about it being best to let sleeping bears lie as we walked back to the kayaks.

On the way back, rain poured from slate-gray clouds as we paddled by the woman's tent. I studied the nylon shelter for movement. The clients seemed to have forgotten her. They talked about the bears they'd seen and asked about the best restaurants in town. The clouds were creeping down close to the ocean when our floatplane appeared out of the gray.

I went to a bookstore and, having little understanding or appreciation for poetry, bought an anthology of Alaskan writing instead. In it were writers who told stories of death, loneliness, and a complicated love for the land. There were stories of the plight of Native people—trauma, resiliency, and the complicated process of learning to adjust to what fate twists lives into. There were stories of hunting and trapping, wolves and bears. I wanted to gift the woman some sense of the place she was visiting.

Not long after, I returned to Windfall Harbor with another group of clients. It was raining, and I kept the book in my drybag where I stored all my emergency and survival gear. We paddled atop black water, studying the shoreline for wildlife. The hair on my neck went up as we neared Middle Creek. There was no sign of the woman's tent. I told my clients about the situation, then beached my boat, pulled out my shotgun, and walked up to where she'd been camped. Hundreds of salmon splashed in the stream, but there was little bear sign. The grass was smothered down to the dirt inside the perimeter of where her electric fence had been. It looked as if she'd imprisoned herself within the small bit of technology, hoping it would hold back the wilderness. I listened for a few moments before returning to my kayak and paddling out to join the group.

"Where did she go?" one client asked, staring wide-eyed at the salmon stream and mist-shrouded mountain.

"Hopefully home," I said.

Flocks of gulls fought over salmon carcasses. I studied the rainforest and, for a moment, entertained a romantic notion. Maybe the woman had crossed into another world. Maybe she threw her tent, electric fence, and gear into the ocean. Maybe she walked into the woods and married a bear.

A short time later I found out that the woman, tired of the rain and the way the bears were acting, had gotten a twenty-mile floatplane ride to Mole Harbor. There, she paddled and portaged a series of lakes, the same Hasselborg once haunted, to Mitchell Bay on the west side of the island. She had not asked about the bay and had no understanding of the dangerous tidal variations in its narrow passageways. A friend of mine rescued her and brought her to Angoon, where she was able to get on a floatplane and leave Admiralty Island for good.

A month and a half after her departure, in mid-October, a passing boat found a small skiff adrift in Poison Cove on Chichagof Island, twenty-five miles away from Middle Creek. The boaters landed on the shore and were confronted by an aggressive brown bear and her cubs. The Good Samaritans radioed the Coast Guard, and soon Alaska State Troopers, Sitka Police, and Sitka Mountain Rescue arrived. A camp and evidence of a struggle were found. Beaten-down and torn-up vegetation led to a bear cache of a mostly eaten man. The consensus was that after having boat problems and being forced to make an impromptu camp, the victim, Tomas Puerta, a veteran Forest Service employee and experienced woodsman, had been attacked by a predatory bear or bears. No one had been killed by a bear in northern Southeast Alaska since 1988, when a man hunting deer in early November was attacked on Baranof Island, dragged up a mountain, and partially eaten.

THE FOLLOWING SUMMER I SAW HARDLY ANY BEARS AT MIDDLE CREEK—perhaps they were still uneasy because of the woman. One of the few bears I encountered there behaved in a way I'll never forget. It was early July, just as pink salmon were beginning to flood into creeks to spawn. MC and I had taken off half a week to kayak, hike, and camp on the island. In Swan Cove, a large solitary male killer whale with a hooked fin approached us. It rose and sank in place, forty yards from the shore, numerous times. In the heat of the day, we watched a female bear with two cubs floating and splashing in the ocean to cool off. We made a sweaty hike up a mountain, along the way surprising a small bear. A pair of merlins—a kind of small falcon—watched us as we rested in the shade and swatted at biting flies. That evening we saw two bears cruising the beach. An adult male near Middle Creek walked down to the water's edge not far from where we floated in our kayak. He was skinnier than he ought to have been, his muscles rippling through his summer's short fur. The bear lowered his head, flexed his forearms and shoulders, and exhaled loudly, letting us know he wanted us gone.

The following morning MC picked salmonberries a few hundred yards from the mouth of Middle Creek. I was standing nearby, watching salmon leap from the calm ocean, when I heard a bear moving through the nearby maze of branches and thorns toward us. I spoke to it gently. The brush crackled, and then a tense silence ensued. I apologized for our intrusion, and MC and I backed away. While we walked toward the Windfall Creek estuary, I wondered if it was the same male that had postured at us the previous evening. I looked over my shoulder to make sure the bear wasn't trailing us. The Windfall Creek estuary was crisscrossed with bear trails, beds, and salmon carcasses. Most of the bears were napping in dense cover or fishing in the shady ripples deeper in the rainforest. Eagles and ravens were waiting for the day to cool off before feasting on the leftovers. The only stirring, besides the salmon constantly splashing, was a flock of noisy Bonaparte's seagulls at the mouth of the stream. We sat on a log hoping to see a bear prowling in the shadows where the stream disappeared into the woods. MC read a book while I listened and watched as the sun fell lower on the horizon. In a few hours, when it turned to dusk, bears would emerge to fish the estuary.

We had a two-hour paddle back to camp, where we were looking forward to a fire and dinner, so we walked back to our kayak. I called out a warning as we approached the edge of the woods, then waited and listened for several moments before approaching the boat. Next to its plastic hull lay a silver salmon with a single large bite taken out of it. I wondered where the bear had caught such a silver salmon—all the fish we'd seen in streams had spawning coloration, and the nearest salmon stream was over a quarter mile away. I did not know what the bear was trying to convey to us. Had it happened upon our kayak and dropped its fish out of fear? Or maybe leaving the salmon was a territorial display of dominance—it wanted us to see the radius of its bite. Maybe the gesture meant something entirely different. Suddenly, I sensed the bear was bedded just yards away. I spoke to it gently, and a branch snapped as the bear turned to the sound of my voice. Then there was silence. The tension became almost audible.

"Let's get out of here," MC whispered.

"We're sorry. We're going now," I said to the bear and hauled the boat to the water.

I LEFT MIDDLE CREEK THINKING ABOUT THE STORY OF THE WOMAN WHO married a bear. The narrative likely originated more than 10,000 years ago and was carried with people as they migrated to America. Variations of it are told across the Northern Hemisphere. I'd heard Tlingit versions of the story while growing up in Juneau, but it wasn't until the summer I encountered the woman at Middle Creek that I'd given it much thought. There are different tellings, but the basic gist is that a woman, while out picking berries, says something insulting about bears. Soon, a handsome man appears, woos the woman, and takes her to his home to be his wife. Sometime later, the woman begins to realize there's something strange about the man and her new home. She increasingly misses her village and the family she left behind. In the late winter, she births two children. Gradually, she comes to realize her home is a den, her children are cubs, and her husband is a bear. The woman's brothers come hunting for the bear. The bear and all but the youngest brother die in the encounter. The woman—under instruction given by her husband before his death—tells her brother the proper funeral ritual he must give the bear. The young man listens to his sister and honors the bear's spirit, avoiding the wrath of other bears. Some stories have the woman and her bear children dying at the arrows and spears of people that were her kin. Others tell of the woman, unable to live a human life anymore, spending the rest of her existence wandering the woods alone as a bear.

There are many layers and teaching points to a Tlingit story, but what I take away most from the narrative is that a line exists between the bear and human worlds. Attempting to cross it can lead to violence and death. There have been other people in recent times, besides the woman camped at Middle Creek, who have tried to bridge our two worlds. The three most famous cases, Timothy Treadwell, Vitaly Nikolayenko, and Charles Russell, ended with bears and people dying. Russell, who was on the Kamchatka Peninsula in Russia, was the only one of the three men not killed by a bear, but he'd habituated numerous subadult and female bears to trust people, and those were killed by poachers.

I know that bears do not want me in their world. Sometimes when I forget my fear and common sense, though, I question the line that separates me from the bear world. Sometimes—and there have been opportunities during my career as a bear viewing guide—I wonder what it would be like to reach across and touch the living animal.

One April, while camped in the central Brooks Range, I heard what sounded like an old woman sobbing outside my tent. Her hoarse cries indicated she was slowly moving across the land. I lay half-awake, feeling powerless and afraid but unable to resist peering outside. Only shadows moved across the snowy tundra and mountains in the star-speckled darkness. The following morning, I found the tracks of four grizzlies—two were alone and the other set belonged to a mom and her yearling cub—all of which appeared to have emerged from their winter dens on the same mountain during the night. I scanned the open space, studying the trails disappearing into the expanse. Maybe it had been her. Maybe my smell and the sight of my tent had stirred ancient, painful memories. Maybe she cries when she wakes from her winter sleep because it's then that she most clearly recalls what's happened to her. I studied the white landscape for a while longer, hoping to see a bear plodding through the snow.

CHAPTER 13
Meat

*The Grizzly Maze, Timothy Treadwell had named this place where he died:
a labyrinth of tunneled trails bears had worn through the dense brush over
centuries of passing—dim, narrow passages where a person could only go
by crawling on all fours, with no telling what lay around each bend, and no
retreat possible. It's a place we've all been, if only in dreams. A fitting meta-
phor for the maze humans and bears have wandered together across time,
and for the story of Timothy Treadwell, which already seemed to defy the sure
paths of logic.*

—Nick Jans, *The Grizzly Maze*

On a night in early August, I lay in my tent on a mountain on Admiralty
Island reading Nick Jans's *The Grizzly Maze* by headlamp. The book is
about Timothy Treadwell and his obsession with brown bears that ultimately led to him and his companion Amie Huguenard getting eaten in
Katmai National Park & Preserve on the Alaska Peninsula. The double
fatality—quadruple if you include the two bears officials killed when they
went to investigate Treadwell's ransacked camp—is the only incident of
a bear killing people in the century since Katmai became a monument
and then a national park. It's also the most well-known fatal bear mauling in history. One hundred yards from where I lay, the meat and bones
of a buck I'd shot a few hours earlier hung from a tree. I could smell the
deer's blood and musk on me. When I got to the part in Jans's book that
describes the gory details of the attack, I had to put it down. I rested

my hand on my rifle, listened to the wind and rain in the darkness, and thought about being meat.

My family mostly hunts deer, but my older brother, Luke, used to hunt black bears. Sometimes I would help him skin and flesh a hide and butcher the meat. I remember one large bear and how we crouched over his skinned body.

"He looks so human," Luke said and then began to sever a quarter from its pelvis.

Luke stopped hunting bears a few years later. The meat of the last bear he shot tasted wrong—it had probably been feasting on the rotten flesh of another bear or a winter-killed mountain goat. It was more than that for Luke though. As a bear dies, it often moans or cries like a person, and if you walk up to it in time, it might lock eyes with you—which is something I've never had a deer, moose, goat, or caribou do.

I've only eaten brown bear once. A friend gave me a backstrap of a ten-year-old male shot in May on Admiralty Island. I had cooked some black bear a few nights prior, but the dark purplish flesh of brown bear looked, smelled, and felt different. The muscle was woven together in strands of flesh that seemed alive. When I laid a pound of flesh on a cutting board, sliced it thin, and doused it with seasoning, I couldn't shake the feeling that this was the closest I would come to eating human flesh.

My night camping alone on Admiralty passed quietly, and at home the following evening, I butchered the deer into steaks and roasts and then ground lesser-quality portions into burger. In all, the buck yielded about sixty pounds of meat. His bones would make about three quarts of broth. I loaded fat, bloodshot meat, and testicles into my backpack to return to the forest. A stiff southeasterly set the dark, wet woods of Douglas Island whispering. My dog, Fen, trotted ahead as we walked to the base of a mountain. I thought about a recent bear viewing client off a cruise ship who had asked if I was a hunter. I'd nodded and, to be polite, asked the man if he was too. He'd told me he was an animal lover, that hunting was barbaric and using a rifle was cheating. He compared hunting deer to shooting cows. I'd shrugged and mumbled something about loving animals too.

I walked off the main trail and into a series of meadows. Rain clouds shrouded all but the lower ramparts of the mountain. It had been such a dry summer that I was happy for the deluge. Salmon were finally able to move upstream and give their spawn and flesh to the woods. I could smell the deer I'd killed. My hands, even after several washings, would smell like its flesh for at least the day to come. Fen ran circles around me as I walked across a muskeg. I pushed through berry bushes to a fringe of conifers. I chose a twisted hemlock to leave the deer remains at the base of, and inadvertently put a foot through a hole surrounded by roots. It seemed like as good a place as any, so I dumped the scraps in and said my last apology and thanks.

For dinner, I sliced the deer's heart into thin strips that I brushed with sesame oil, turmeric, and salty seasoning. I cooked it on a cast-iron skillet then sat down for my meal and watched out my window as waterfalls cascaded down the gray mountainside and a Disney cruise ship motored through Gastineau Channel blasting the corporation's theme music. A couple thousand passengers and a similar number of crew members were aboard the ship. I imagined all those people sitting down to all-you-can-eat buffets. The last notes of the Disney song faded, and I took a bite of deer heart. I thought about how much easier it is to play make-believe rather than acknowledge all the violence and death that comprises each of our lives.

A FEW DAYS LATER, I LEFT TOWN TO SCOUT A VALLEY FOR A FILM CREW that wanted sequences of wolves catching salmon at night. I checked different spots on streams and sloughs until I came across a young gray wolf. It watched me for a few moments before retreating to the brush. Wolves had been fishing nearby riffles, filled with chum and pink salmon.

On a gravel bar were numerous headless salmon—wolves only seem to eat the head, probably to avoid parasites that live in other parts of the fish and affect them more negatively than they do bears. On my hike back, I found the fresh tracks of two brown bears—one a big male and the other a smaller animal. While I waited for the crew to arrive, I finished Jans's *The Grizzly Maze* and began a book called *Among Grizzlies: Living with Wild Bears in Alaska*, which Timothy Treadwell, with the help of his friend Jewel Palovak, had written and published in 1997.

I first heard about Treadwell shortly after his death in 2003, when a trapper from the village of Eagle picked me up hitchhiking. He told me two Californians had been eaten by a bear on the Alaska Peninsula, that their camera had been on and recorded the audio of the attack. In Fairbanks, where the trapper dropped me off, I read the transcript of the audio and then wished I hadn't. A few years later, the documentary *Grizzly Man* by famous film director Werner Herzog was released. In true Herzog fashion, the film was mesmerizing and impactful. In it, you see a tortured soul who finds meaning in being close to brown bears—so close that he touches, even kisses, the bears at times. In the past, Herzog has admitted he's more interested in telling stories depicting a poetic truth versus the actual truth. Nick Jans, who spent a year devoted to Treadwell while writing *The Grizzly Maze*, was shocked by how much Herzog's portrait differed from his findings. Jans wrote he had expected a portrayal with a positive bias:

> Instead, in a montage woven from Timothy's own self-shot videotape from Katmai interspersed with staged scenes and didactic narrative, *Grizzly Man* presented a disturbing vision of a man teetering on the edge of sanity, then plunging into self-destructive madness as his world unraveled—essentially committing suicide by bear. . . . Missing almost entirely from the Herzog film was the other side of Timothy—the charismatic, heady showman and altruistic con who hobnobbed with stars and became a celebrity, the Timothy so many people described as affable, quick-witted, and charming.

When Treadwell was in his early twenties, he fled his life on the East Coast and moved to Los Angeles to reinvent himself. There, he joined multitudes of other young people who wanted to escape their past and become famous. Instead, Treadwell spiraled into darkness, drugs, and alcohol. In 1989, he fled again, this time to Alaska. He discovered brown bears and was instantly enthralled. He wrote in *Among Grizzlies* about his first close encounter with a bear in Katmai National Park & Preserve, a region on the Alaska Peninsula renowned for its easygoing bears. Katmai is closed to hunting, has prolific salmon runs, and has an incredibly dense population of bears, which all adds up to make it the best place in Alaska for people to be close to and watch bears. Treadwell's first encounter with a Katmai bear likely occurred near Brooks Camp, which currently attracts upward of 350 bear viewers each day during the salmon runs. Treadwell was walking on a trail through thick brush when a bear appeared, slowly coming in his direction. He wrote:

> The bear quickly closed the remaining distance between us. I stumbled on the wet mud, falling down hard, face first. As I curled into the fetal position, the grizzly's steps vibrated the ground right up to my sniveling face. Pearl-dagger claws stopped inches from my cheek. Spreading my fingers over my eyes in a bizarre form of peekaboo, I gaped up at the grizzly's face. What little light remained exposed an immense furry face, horribly engraved with long, deep scars that told of past battles. He exhaled a puff of fish-scented breath, then quickly inhaled my odor. No brilliant defensive strategies occurred to me. I was on overload with terror, and simply seized up. The grizzly hovered above for a few minutes, but for me, down on the ground, it felt like a lifetime. Then, ever so gently, he stepped over my quivering body, his bloated tummy scraping across my right shoulder, and vanished in the direction of the river.
>
> Long after he was gone, I picked up my shaking body and clumsily left the woods, all the while chanting, "Thank you, bears . . . thank you, bears."

After that encounter, Treadwell began spending summers and falls living among the bears of Katmai. During the winter he returned to California and promoted the "work" he was doing with brown bears. Similar to Doug Peacock, Treadwell credited grizzlies with saving his life. Both men shared a feeling of reciprocity and a desire to protect the species, but their approaches to conservation and interacting with the animal were very different. Peacock appreciates bears for what they are—the fact that they are predators is something he celebrates. They're not safe, nor does he want them to be. He recognizes that bears need space, not just from civilization but from him as well, and he goes to great lengths to avoid stressing and displacing animals. Peacock's work, centering around the preservation of grizzlies in the Lower 48, reflects a philosophy of restraint, humility, and respect.

Treadwell, on the other hand, was promoting an idea of bears as something out of a Disney movie. Herzog believed Treadwell saw his life akin to a movie and that he took the role of director and lead star as he treated the bears like supporting actors. Despite being situated in a national park that was heavily patrolled, he claimed to be the only thing protecting the bears from a supposed epidemic of poaching. Most of the rest of Alaska's bear population is subjected to hunting—in many areas the leading cause of death for adult bears is a hunter's bullet. If Treadwell really wanted to save bears, he could have gone to Karluk Lake on Kodiak Island, where the biggest bears in the world have been shot, or Hood Bay on Admiralty Island, where numerous bears are killed in the spring and fall, just to suggest a couple places. Hollywood dreams die hard though. Katmai bears made the ideal subjects around which Treadwell could spin his self-aggrandizing myth. They're still brown bears, and there's no denying Treadwell's raw courage (some would call it stupidity). The fact that he stayed alive for so long doing what he did is mostly a testament to how tolerant brown bears can be. But, in the end, due to his high-profile death, he ended up reaffirming negative beliefs about bears and, likely, doing more harm than good for the animal.

Treadwell's claim that he was working to protect bears resonated strongly with many people, but his narrative got more bizarre and enigmatic from there. Joel Bennett, a Juneau wildlife filmmaker, remembers encountering Treadwell in Seymour Canal on Admiralty Island. It was at the start of Treadwell's career with bears, and he was a mess. He couldn't walk down the beach without falling, set up his tent, or do other backcountry basics. Despite Treadwell being an easy target to judge or dislike, Bennett befriended him, and the two worked together filming bears in Katmai and on Kodiak Island numerous times. Treadwell didn't spend much time in Southeast Alaska. He likely found the bears of Admiralty Island too elusive. Except for bears at Pack Creek, ABC Island bears generally behave very differently than Katmai bears by going to great lengths to avoid people. Treadwell would end up spending thirteen seasons camping with and filming bears up close and personal around Katmai's Hallo Bay and nearby watersheds. Fifteen years after Treadwell's death, and a tidal wave of criticism, Bennett still quietly defended his friend.

"He wasn't crazy. He knew a lot about bears. I always felt safe with him. There were certain bears he knew to avoid. He wasn't reckless," Bennett said.

There are a lot of people who believe the opposite—though none spent the amount of time with Treadwell that Bennett did. Many "bear people" and Natives considered his in-your-face way of interacting with bears beyond disrespectful. He camped in alder thickets laced with bear trails, constantly surrounded by bears. These are places I would avoid walking through, let alone camping in. On numerous occasions, bears threatened his life, destroyed his camp, and pushed him to the brink. Once, wildlife filmmaker Mark Emery was flying in a floatplane and saw Treadwell standing in the ocean surf yelling frantically and waving his arms. They made a difficult landing and taxied over to Treadwell, who was hysterical about a bear that had just shredded his tent, in which, he said he'd been storing dog food to feed foxes. The interaction became more bizarre when Treadwell, noticing a woman from the film crew approaching, switched to speaking in an Australian accent. He nudged Emery and whispered, "Women love that Australian accent."

Treadwell relied heavily on the kindness and generosity of bear viewing guides, lodge owners, pilots, and others. He usually left Alaska by September or October and spent the rest of the year in Malibu or elsewhere in the Lower 48. There are stories of him showing up to speaking engagements in a suit several sizes too big to play the part of a clown and, at least at one event for adults, doing his entire gig using sock puppets. Shortly before his death, Treadwell began to achieve the fame he had long hungered for. Most Alaskans called bullshit, but Treadwell's narrative captivated many people out of state, including well-meaning celebrities like Leonardo DiCaprio and environmental brands like Patagonia. People who knew little about brown bears saw footage of Treadwell cooing to bears placidly eating grass just feet away and ate it up like a Disney fairy tale. Often Treadwell would be talking in the film, saying something like "This is my friend Mr. Chocolate. He saved my life, and now I work tirelessly to save his," in a high-pitched singsong voice, as a hulking male bear grazes in the background of the shot.

THAT AUGUST WHILE I WAS GUIDING THE WOLF FILM CREW, I HAD TO STOP reading Treadwell's book, *Among Grizzlies*. Spending nights sitting alongside a salmon stream, waiting for a brown bear to come in the form of a shadow in the darkness, and having his voice in my head had a way of inspiring saner reading. The shoot couldn't have gone better. We displaced only one bear. Wolves came within ten yards of us during the blackness of night. The camera operator, using a military night camera, got footage of wolves catching pink salmon. That morning, just as mountains began to appear in the first light of dawn, we packed up the heavy kit and began the hike to camp. At one point we paused to look back and were surprised to see a female wolf trailing us just yards away. We froze, and she came closer and then lay down and rested for a few minutes before walking back into the woods.

In July of 2003, Joel Bennett met Treadwell on Kodiak Island to work on a short film for Disney about Treadwell's life with bears. The plan was that the short would run at the beginning of the corporation's star-filled animated film *Brother Bear*. Treadwell had finally made good on his life-long Hollywood dream. He was at the top of his game, and the future was rife with possibility and stardom.

After the Disney shoot, Treadwell flew to Upper Kaflia Lake in Katmai and spent six weeks brushing shoulders with multitudes of bears as they feasted on salmon and tried to put on enough weight to survive the upcoming winter sleep. September came, and the bears entered hyperphagia. Other seasoned bear photographers, noting aggressive changes and environmental factors further stressing the bears, packed up and left. Treadwell's journals show he was aware of heightened aggression but he decided to not only stay past mid-September but also have Amie Huguenard fly out and join him.

It's unclear what Huguenard was to Treadwell. By most accounts she was terrified of bears and did not really want to be camped in what Treadwell called "the grizzly maze," especially so late in the season when the weather was at its worst and the bears their most dangerous. Their time together was stressful, and it must have come as a big relief for Huguenard when she heard the drone of a floatplane approaching on September 26. In a day or so, she'd be back in the sunshine, smog, and bustle of LA. The couple flew to Kodiak, stashed gear at a house owned by a friend of Treadwell's, and headed to the airport to fly south. At the Alaska Airlines counter, an agent told Treadwell what it would cost to change his ticket, which was scheduled for a later date. Treadwell stormed out, and three days later, he and Huguenard were back at Upper Kaflia Lake camped amidst bears—many of which Treadwell didn't know, despite his seasons there—at the peak of hyperphagia.

Around two o'clock on the afternoon of October 5, a day before Tread-well and Huguenard were scheduled to return to Kodiak, they heard a bear outside their tent. It was hardly unusual, considering their camp spot, but this bear was different. Treadwell exited the vestibule to confront the bear—something he'd done a multitude of times—while he or Huguenard flipped on the video camera to get the encounter on film. Treadwell tried talking and a variety of physical displays, but the bear, instead of calming or retreating, came for him. The video camera ran for six minutes, recording Treadwell's screams, pleas for help, and after about four minutes, his silence. It's unclear at what point the bear began feeding on Treadwell—it could have been as soon as it knocked him down or sometime later. The tape lasted two more minutes, recording Hugue-nard's high-pitched, hysterical screams. Sometime after the tape ended, the bear came for her. Her mostly eaten remains were found buried next to the flattened tent.

The theory is that it was an old male bear that was responsible for killing Treadwell and Huguenard. When authorities went to investigate the camp, the bear was guarding their mostly eaten bodies. They killed it and later, when a subadult made an aggressive display, shot that bear as well. It's interesting to note that the older bear had been tranquil-ized and collared years before, an invasive process that some people hypothesize increases a bear's aggression toward humans. A necropsy revealed that its stomach was filled with Treadwell's and Huguenard's flesh, bones, and hair.

Nick Jans flew to Upper Kaflia Lake to investigate Treadwell's campsite just days after the fatal mauling. As he was walking through the alders, he could smell Treadwell, Huguenard, and the two bears the officials had killed. Jans, like any relatively sane human, confessed he'd "been scared shitless." It didn't help when a large male bear, who'd likely been feeding on the remains of the two bears killed by officials, came out of the brush and escorted him back to the lake.

I FINISHED TREADWELL'S *AMONG GRIZZLIES* THAT AUGUST WHILE GUIDING bear viewing trips to Admiralty Island and Chichagof Island. My last trip was to Pack Creek, with two youngish couples. One man and woman were vegans from San Francisco—they were gregarious and likable, and they had dreamed of visiting Alaska for years. The main reason for their trip, the woman confided, was that her husband "wanted to look into the eyes of a brown bear and have a moment." The bear viewing was good that day, but the Californians seemed to want more. At one point after we walked the mile-long trail through the woods leading to a twenty-foot-tall tower above salmon spawning beds on Pack Creek, a dark female bear paused from fishing and stared up at us. It was an unusual display, and her small, dark eyes radiated a bored contempt. After she continued on her way, the vegan man had a strained, almost sad look on his face. Later that day, a massive bear I did not recognize appeared downstream, and my skin crawled with electricity. Its presence at Pack Creek meant that hyperphagia season had begun. Other bears that hadn't been using the watershed during the summer would begin transiting through. I directed everyone's attention to the giant bear, but at one hundred yards it was too far away for their interest. A few moments later the bear disappeared into the dark rainforest.

There were still trips to guide, but I needed space from people who wanted things from bears and me that neither of us were capable of fulfilling. I spent the first half of September recalibrating and hunting to put up a supply of meat for the year—something I'd done nearly every fall since I was a teenager. I severed the hearts of animals from their chests, cut their quarters free, and carved the flesh from their bones. Once, when I was knelt over a deer I was butchering, my mind conjured up an image of a crouched bear eating Treadwell. I imagined him screaming and moaning as the bear tore his flesh and organs from his viscera. Then I imagined a bear crouched over me, tearing and gobbling, and I wondered if there would be anything other than horror in those final moments. I studied the blood, guts, and flesh of the deer, then cupped the buck's face and uttered a quiet prayer.

CHAPTER 14

Wilderness of Brown Bears

> *Not merely yarns or episodic adventure narratives, bear stories speak to the timeless battle between the human race and nature, and to the bear's singular status as nature's supreme symbol in the Northern Hemisphere. The bear is not only the nemesis of the stockman, beekeeper, and camper, he is the preeminent carnivore, the local embodiment of terra incognita, the visible reminder of the disturbing powers that defy and outlast us.*
> —David E. Brown and John A. Murray, *The Last Grizzly*

In April of 2016, I met my friend Forest Wagner for coffee. We'd gotten to know each other sixteen years prior in an anthropology of religion class at the University of Alaska Southeast. Our friendship, based on a love for all things wild, grew from there. Each spring we tried to have a check-in before we each got too busy with work and adventures. That year, we joked about getting older and talked about past adventures and trips we hoped to take some day. Forest was departing the next morning to lead a college mountaineering class on an attempt to climb Mount Emmerich above the Chilkat River. Later that week, I was planning to leave to walk the length of Admiralty Island with my brother Luke. We hugged goodbye, told each other to be safe, and planned to meet up in late summer when Forest was back from fishing Bristol Bay.

A few days passed before I got a phone call from a friend with ties to the university. He told me Forest had been attacked by a brown bear, medevaced to Anchorage, and was in critical condition. I postponed leaving for Admiralty and flew north. Forest looked like hell and was hooked up to numerous machines, but as nurses worked on him, he smiled and asked, "Do you want to see my wounds?"

Two days later, I stared out the window of a floatplane as slate-gray clouds hid the mountains of Admiralty. I studied the white froth of waves in the roiling black ocean as Luke, sitting behind me, tried to keep from vomiting. A landing in Murder Cove, on the southern tip, would be impossible. The pilot gave me a nervous look. I yelled over the roar of the plane's engine and asked him to try landing in Whitewater Bay, the southernmost protected spot on the western side of the island. After circling the bay several times, he set down in the heavy winds and whitecapping waves. Luke and I hopped onto the shore, and a few rushed minutes later, the plane took off and disappeared into the gray, taking our bear spray with it.

"Shit!" Luke said. "We may have needed that."

Fresh bear tracks cut across a stretch of sand as sheets of rain undulated from grim layers of clouds.

When I'd decided to walk the length of Admiralty, a journey I guessed would be roughly two hundred miles long with all the bays and inlets, Luke was one of the few people who didn't think it was a crazy idea. A family man, not prone to wandering unless it involved hunting, he surprised me when he decided to join my walk. After what happened to Forest, I'm not sure I would have had the courage to do it alone, or at least to stop thinking of one of Hosea Sarber's sayings—that one out of twenty-five brown bears you meet will want to fight you, often for reasons known only to the animal. There are some bears, whether starving, wounded by man, deranged, or simply having a bad day, that are difficult to reason with. I loaded my .44 Magnum, then shouldered my backpack, and Luke and I began trudging north along the rocky beach.

The first bear we encountered had been killed by hunters. All that remained was a skeleton; the paws and head had been sawed off and the rest picked clean by eagles, other bears, and sand fleas. An hour later, we came across the pawless, headless skeleton of a larger bear. At last light, in a small cove, a female with two spring cubs stared at us before running into the woods. Caught in the dark, we were forced to camp nearby. We lay in the tent listening to moaning and screaming, not 100 percent sure if we were hearing a cub that had gotten separated or the wind.

In the early light, Luke and I ate breakfast near a grandfather trail we had accidentally camped next to. We traveled on through rain and wind, mostly keeping to the rocky shore. We jumped a bear while hiking through the dense forest to Hood Bay; the brush exploded as it ran away, leaving us in an eerie silence. On the beach we came around a corner and displaced a skinny bear cruising the high tide line. A half dozen eagles, feasting on a giant beached skate, took flight as we approached. We inflated packrafts near where the north and south arms of the bay met and made a windy half-mile crossing. I studied the dark blur of South Arm Hood Bay and thought of our family friend Paul Kissner, who'd been mauled here in 1967.

I HAVE A VAGUE CHILDHOOD MEMORY OF SEEING A BROWN BEAR MOVE through salmonberry bushes along a stream bank on Admiralty Island, perhaps real or imagined, but the first time I know for sure that I saw a living brown bear was on a camping trip with Paul. It was near a commonly used trail on the road system north of Juneau. With the amount of human traffic, it was uncommon for brown bears to use the area. I remember seeing a distant brown animal and looking at Paul, who was carrying a big pistol. We knew Paul had been mauled, and we also knew we weren't allowed to ask questions or talk about it in his presence. That brown bear remains a blurry dot in my memory. Now I would guess that it was a subadult that had wandered in from the wilder country to the north. It was killed not long after—a fate nearly all brown bears share when they cross the line from their world into ours.

Fifty years after Paul was mauled, I sat down and had a cup of coffee with him. He'd told few people the story, though an inaccurate version of the events had been printed in Larry Kaniut's *More Alaska Bear Tales*. Despite the passing of half a century, it was still emotional for him to talk about. He was twenty-one when it happened, fresh to Alaska and beginning a job for the Alaska Department of Fish and Game as a fishery technician in South Arm Hood Bay. On July 19, 1967, Paul and his partner, Bruce Milenbach, were going to go up to Falls Creek to survey residential Dolly Varden above a waterfall. The salmon had yet to begin running, and bears were feeding on the profusion of vegetation in the high country. When Paul and Milenbach approached the subalpine, at an elevation between 1,500 and 2,000 feet, they began to see a fair amount of bear sign. Paul, who was new to brown bears, was in the lead when the two men came to a cliff.

He was about to blunder into what he believes was a mom with cubs, triggering her to attack out of fear over the safety of her cubs. Male bears, especially during the spring mating season, will pursue females with cubs. The stress this dynamic causes a female bear can sometimes be displaced onto a human, especially in a surprise situation where she has little time to reason whether she's facing a threat.

"I took my pack off," Paul said, "which had a .44 Magnum pistol in it, and climbed a vertical rock face about twenty feet and then walked through a patch of alders. All of a sudden in front of me, something roared. I turned around and yelled back a hundred yards or so to my buddy, 'Shoot the gun! Shoot the gun!'" He was hoping it would scare the bear away. Paul continued: "I stepped behind a big alder and could hear a bear snorting and the brush crunching. The bear came right through the middle of the alder and grabbed me by the thigh and just picked me up and shook me. I put my hands on the inside of its jaw to try to do something."

When the bear set Paul down, he moved backward and fell over a twenty-foot cliff and landed on a ledge. The bear leapt down and bit Paul's waist and upper hip. When the bear let go, Paul rolled off the cliff and tumbled three hundred feet down an avalanche chute. The bear ran after him until it lost control and it, too, rolled down the mountain. Paul and the bear careened over about a ten-foot cliff and piled up next to each other. The bear got to its feet, roared, shook its head, and walked away.

"I started to yell for my buddy, but the loudest I could yell was a mere whisper. I was in shock. I could see I was bleeding pretty profusely, so I pulled my belt out of my pants and put it around my leg. I was having a tough time. There was absolutely zero pain. I felt like I'd been hit by a truck. My buddy had a difficult time finding me, and I had a difficult time leading him because I could only whisper."

It was a warm day, so the two men packed Paul's leg in snow. A doctor would later say packing the leg in snow was probably what saved Paul's life. Bruce had no choice but to leave Paul where he lay and go for help. Once he reached the shore, due to the tide, Bruce was forced to swim a hundred yards through the frigid ocean—no small feat in Southeast Alaska, considering the dangerous currents and water temperatures—to the skiff before motoring to a field station to use a two-way radio to contact help in Juneau. It was a miracle that Bruce was able to make the call and get through. Paul shook his head in disbelief as he remembered.

"We used to get reception about two or three days a week . . . and he got back at the field camp at 4:59, and the radio is shut off at 5:00 p.m. every day," Paul said.

A helicopter was summoned, along with a floatplane carrying Dr. Henry Akiyama, a respected physician from Juneau. Paul laughed as he remembered how, after he was lifted via helicopter to the shore, the first thing Dr. Akiyama asked him was if he'd like a shot for pain. Paul answered in the affirmative.

"I was in the hospital for twenty-one days and on workman's comp for about a year," Paul remembered. "Then I went back to Hood Bay after I'd graduated from college. I'd only been back for a month and a half when this second incident happened. It was mid-May and we'd seen hardly any bear sign."

Paul was surveying juvenile salmon. His partner was fifteen feet behind him, and they were walking fairly fast when from less than twenty feet away, a bear rose from its day bed. It roared and then charged. Paul dove to the side, and the bear just missed him. The bear pivoted to charge again when it saw Paul's partner coming up behind. It thought better of it and walked off into the brush. After that incident, Paul radioed to town and went home to think hard about what he wanted to do for a living. A week later, he was back on Admiralty Island.

The irony is that Paul spent the rest of his career surveying streams all over Alaska, deep in the world of brown bears, and never experienced another attack. It wasn't random though. Paul had changed his behavior in bear country after the second bear charge, stressing the importance of walking slowly, making a lot of noise, and being armed. A good partner is equally critical.

"I don't begrudge the bear that got me," he told me. "I was in its territory. I was invading it, and it was just doing what a bear does. I have no animosity. The protection officers wanted to kill the bear and I told them no. That's the way I feel today. Most encounters with bears are brought about by man, by the way we interact with them. Ninety-nine percent of them will turn around and leave. Then two in a row will try to charge you. So, you never know," Paul said, shrugging.

"I had absolutely no thought of bears until I got to Hood Bay," he continued. "Since then, I've studied everything I can about bears. If they really want to get you, they're going to get you. That's the bottom line for me. If they really want you, they're going to get you. They're pretty crafty, and they can be quiet. Most of them don't want you. Ninety-nine percent don't want you."

LUKE AND I HIKED ALONG THE SHORE UNTIL WE ENCOUNTERED A COUPLE dozen eagles fighting and tearing at a pink and bloody hulk. All but the boldest took flight as we trudged near. Finally, the last bird winged to a nearby boulder and stared at us as we stood over the carcass of a bear shot and skinned earlier that day. A seasoned hunting guide once told me that brown bears don't understand death. Luke, one of the most avid hunters I know, grew quiet as we studied the carcass. Water lapped at our boots as the tide crept in and turned the ocean red with blood. Soon, all evidence of this bear, other than a scattered headless and pawless skeleton, would vanish.

That night we camped a few miles away from the Tlingit village of Angoon. We weren't far from the dump, where bears would often feed on rubbish and become conditioned to lose their fear of people. We were happy to break camp before sunrise and hike on. We made a short crossing through stiff currents to the other side of Mitchell Bay, passed a handful of derelict cannery buildings, and were soon traversing a rough, wild shoreline. After hours of hiking through the woods and along cliffy beaches, we encountered a male bear chasing a female with cubs. The latter sprinted into the woods, and the male turned its attention to eating sedge grass. Near dark we stumbled, soaked and shivering, upon the three-walled shack that marked the beginning of a trail and portage across a chain of lakes to the eastern side of the island.

In the chilly morning, we watched a bear graze on the other side of the cove before entering the dripping jungle. Giant spruce and hemlock trees, their boughs adorned with old-man's beard lichen and moss, rose into the fog. Sooty grouse hooted above the trail and the amount of bear sign diminished the deeper we went into the interior of the island. On Davidson Lake, the sun tore through flushed storm clouds, revealing the snowy massif of Thayer Mountain. Late that evening, we portaged from Hasselborg Lake to Lake Alexander and spent the night drying out in a Forest Service cabin. We pushed on through a nasty storm to Mole Harbor on the eastern side of the island. Near Flaw Point, we had to holler repeatedly in the deafening wind at a bear walking our way before it became aware of us. It looked at us, confused for a few moments, before deviating into the woods. The next morning, as we broke camp in heavy rain, I told my brother I'd never seen a bear on the next fifteen miles of shore, though I'd kayaked and hiked along this portion of beach a number of times.

"Then today's probably the day we'll run into trouble," Luke said.

Due to all the heavy rain, the beach was flooded to near biblical proportions. Normally shallow streams had turned into rivers, forcing us to strip off rubber boots and pants before fording the icy currents. Before long we ran into a mom with a spring cub. Then we bumped into a medium-sized bear. A half hour later, as I was putting my pants back on after another river crossing, Luke suddenly jumped.

"Oh, shit!" he yelled, and in one motion I pulled out my pistol and whirled to see a large Shiras bear barreling down on us. Froth and drool hung from the bear's muzzle as it popped its jaw and swaggered back and forth. At thirty yards, the bear paused and shook its head in rage, trying to work up the courage to close the distance. I yelled at the bear, something I rarely do unless I figure there's a good chance contact is about to be made. For two to three minutes, the bear paced the beach in a fury until gradually he appeared to calm a bit. I told him that he needed to be more careful, as there were numerous bear hunters out who would like his hide on their wall. He circled around, trudged up to the beach grass, stood, and rubbed his back on alders before sitting and staring at us. The wind shifted, and when our smell hit him, he looked like he got the scare of his life and charged off into the forest. Leaving cover and seeking out people was abnormal behavior for an Admiralty bear, and it made me wonder if he had a history.

NINE YEARS PRIOR, JOHN RASTER, AN OTOLARYNGOLOGIST (EAR, NOSE, and throat specialist) from Juneau, had been mauled by a bear at nearly the exact same location. Even though a male bear on Admiralty has on average a sixty-square-mile home range, I wondered if it was the same bear. After we returned home, I called Raster up and got his story.

"I actually like bears. I just like that we both give each other space. I'm more scared of people than bears. Bears tend to be pretty predictable, except for every now and then you get that funny one. And that's what happened to me," Raster said.

Raster is a seasoned outdoorsman who has traveled extensively in brown bear country. He's walked many ABC Island salmon streams and looks forward to an annual fall deer hunt at a cabin on Admiralty Island. In 2007, Raster and his friends went out to their cabin a little later than usual. On the last day of the trip, November 30, Raster got up before sunrise to take a walk and photograph a pod of humpback whales he could hear grumbling and whistling in the darkness. He slung a rifle over his shoulder and walked along the frozen beach as the pink-orange glow of sunrise backlit spouting whales. Most bears had denned up for the winter, but one night earlier in the week, the men had heard an agitated bear clacking its jaws outside the cabin. That morning Raster had walked only a few hundred yards down the beach when he heard a loud snap coming from the woods above and a bear came charging out of the black wall of the forest.

"In one and a half seconds, and in three or four big bounds, his nose is two feet above me. He's on his hind legs and ready to pounce on me. I had just enough time to drop my camera. I already had a bullet in the chamber, and I flipped off the safety, and I tried to rotate my shoulder around to fire," Raster said.

He managed to get a shot off but just missed the bear's face. The sound of the bear growling was deafening; the next moment Raster felt like he'd been hit by a car. He rolled down the beach, worried the bear was going to drag him into the woods. The bear pawed and roared at him until he was in the ocean. He thought about trying to swim, but the bear pinned him with its paws and pounded down like it was trying to drown him. The bear grabbed him by the belly and rolled him back onto the beach. Raster finally got a good look at the animal—he guessed it was a four- or five-year-old male that weighed around 550 pounds. The bear went for his neck, and he reflexively tried to protect his spine with his hands.

"I didn't quite get my right hand in, but I got my left hand in before it clamped down on my neck. His lower jaw fang was in my skull and in my neck on my right side. His upper right fang got stuck in my skull, and his left fang got stuck in my Gore-Tex glove. My left hand made it so that it was just big enough that it couldn't get all the way around my neck. He picked me up off the beach, and I hear this loud snap. My thoughts were, *There goes my second cervical vertebrae. Game over*," Raster said.

A moment later, he lost consciousness.

The cold water of the incoming tide woke him. The bear appeared to be gone, but Raster knew there was a strong possibility the bear was still watching and if he moved that it might resume its attack. He lay still, playing dead, until the rising tide turned him hypothermic. Believing there was a strong possibility he was paralyzed, or at the very least had a broken spine, Raster first wiggled his fingers and toes, and then, slowly got to his feet. He walked back to the cabin, resolved to jump in the ocean if the bear returned. The men used a sat phone to call in a helicopter, and before long, Raster was in the emergency room at the hospital in Juneau. They scanned his head, neck, and pelvis and found that the bear's canines had just missed his occipital artery and jugular vein. Surprisingly, Raster's spine was intact. The snapping sound he'd heard when the bear clamped on his neck was actually the long bones in his left hand being crushed.

"The surgeon opened up my hand and put a plate around all of that and then stitched up all of my wounds. I got mauled on a Friday, and I was back to work on a Tuesday. I didn't even miss a day of work," Raster said.

Raster still goes into bear country. He accepts there are risks but feels safer there than in many cities.

I WATCHED THE EDGE OF THE FOREST TO SEE IF THE BEAR WOULD RETURN and thought about how there's always a chance, even when you do everything more or less right, that you'll run into a bear that will want to attack. Traveling alone increases the odds—I wondered what would have happened if Luke had not been with me. Would talking have been enough, or would I or the bear have had to die? Luke and I walked on, passing numerous deer, and paused when another bear appeared, coming our way along the beach. It was medium-sized—probably an adult female. I spoke, and she looked up and, after a moment, ran into the woods.

The rain fell so hard that creeks raged like jet engines. That night as I struggled to get warm in my soaked and rotten-smelling sleeping bag, I thought of Forest lying in his hospital bed hooked up to wound vacs and monitors. I thought of the grizzly that had charged me in the Brooks Range. I had been shaken at the time, but years later I'd come to see the encounter—how the bear had come close to attacking but at the last moment spared me—as a defining moment. Now, as I listened to the incessant rain and wind and thought of my friend, my take on the Brooks Range bear felt cheap and transient. What meaning would I have been able to find if that bear had torn my face off or gutted me?

The next morning, rain clouds swirled over mountain slopes, offering some visibility. We followed a bear trail as wide as a bike path and hiked miles of mudflats to one of my favorite coves. There, we counted seven bears chasing, posturing, and avoiding each other. An old, scary-looking boar pushed a younger male toward us as we tried to ford a swollen creek. The younger male seemed dejected and paralleled us at less than a hundred yards for several minutes without looking our way. At the next crossing, a small bear came trotting and then, when it became aware of us, plunged into the stream and swam across. By the end of the day, my feet were blistered and bruised. Blood from chafing oozed down my legs. Luke's feet were doing better than mine, but his hands were shriveled, blistered, and on the verge of turning into rotten mush from being perpetually wet and cold. That night at camp it took gun oil and a lot of pulling to get his wedding band off his swollen finger.

At first light we watched a large bear walking along the edge of the trees. When it reached the border of our camp, it huffed and ran into the woods. We paddled the last five miles to the head end of Seymour Canal, where we hoped to use a cabin to dry out and give our feet a break, but it was occupied by bear hunters. When we emerged from Oliver Inlet, two deer swam to shore and a humpback whale exhaled, echoing across the still waters of Stephens Passage. The sun burned through clouds, and for the first time in days, the rain ceased. Multiple bears and deer fled into the woods as we hobbled along the rocky beach. At Young Point we set up camp as the setting sun lit the snowy Chilkat Range crimson and gold. Loons sang eerily, and a humpback whale let out a whooshing breath.

"Will you model for a photograph?" I asked Luke in a moment of levity. He pretended to walk down a fashion runway and then struck a pose on a log. Our laughter was cut short by a nearby bear running away into the brush.

On our last day, we began skirting around Young Bay before sunrise. A giant cruise ship, glowing with gaudy carnivalesque lights, lit up the ocean. At the dock for the Greens Creek mine, built outside wilderness boundaries and one of the largest silver operations in the world, two workers asked to take our picture. Lodges and cabins appeared on Horse and Colt Islands, which are separated from Admiralty by only a few hundred yards of ocean. Four miles away, the buildings and docks of Juneau dotted the mainland. A stiff wind stirred the seas into a froth, and rain began to fall. We were a couple miles from the northern tip of Admiralty, but it was time to go home.

In a few hours, we'd be safe with our families. I'd call Forest, and he would tell me about some of the future wild trips he was planning. I would cringe but offer encouragement. It was hard to imagine him walking a mile, let alone climbing mountains again. Right then though, I studied the tracks of a large bear leading into the dripping, swaying forest. Dad's skiff appeared on the rough ocean, and he cautiously made his way through a rocky passage. Luke and I waded out into the ocean to keep the boat from being thrown up on the shore and then leapt up onto the bow. As we motored home, I looked back at the dark mountains of Admiralty disappearing into slate-gray clouds.

CHAPTER 15

The Trail to the Mountains

Gradually, perhaps over many centuries, the human question went beyond "How do we survive the cold winter?" to "How do we survive the cold death?"

The bear, more than any other teacher, gave an answer to the ultimate question—an astonishing, astounding, improbable answer, enacted rather than revealed. Its passage into the earth, winter's death, and burial under the snow was like a punctuation in the round of life that would begin again with its emergence in the spring. . . . The bear seems to die, or to mimic death, and in that mimicry is the suggestion of a performance, a behavior intended to communicate.

—Paul Shepard and Barry Sanders, *The Sacred Paw*

Throughout the night we listened to brown bears popping open the skulls of salmon. It was mid-September and the end of the 2016 bear viewing season.

"Wow!" my boss and friend, wilderness guiding legend Ken Leghorn whispered dreamily as he lay in his sleeping bag. We could hear the sounds of salmon thrashing water to dig redds and bears prowling over the rush of the stream. I stared out into the darkness, studying the shifting shadows and listening.

In the morning Ken sat in an estuary while our friend Dan Kirkwood and I attempted to make plaster casts of bear tracks in the mud. The mountains of Admiralty Island rose into clouds. Eagles and ravens perched on the branches of ancient spruce and hemlock trees. An occasional bear would appear and check the stream for fish before receding back into the forest. A female bear walked out into the meadow, lay down, and nursed her cubs of the year. This trip was Ken's way of having us say goodbye to another season of guiding bear viewing trips. When we walked to meet our floatplane, an old, scarred, crippled bear—who'd been given the name Patches by Pack Creek rangers—blocked our way for a few minutes before slowly limping away into the forest.

A few weeks later, Ken called to tell me he'd been diagnosed with stage IV pancreatic cancer. He ended the conversation by thanking me for going with him on the camping trip to Admiralty. It was not long before he began having difficulty digesting. Doctors told him ground wild game would be some of the best food for him, so I promised to get a deer and grind it into burger for him.

I stood at the edge of the woods on Douglas Island a few days later, listening to the wind make the forest creak and groan. I loaded my rifle's magazine, then pushed through brush and began following a deer trail toward a mountain. Old-growth trees rose just beyond the yellow grass and shore pines of the muskeg. Fresh tracks appeared, and I tried to focus, but my thoughts kept going back to my friend. I'd heard a lot about Ken before I went to work for him. He was considered the father of ecotourism in Southeast Alaska, and experienced wilderness guides spoke reverently of him. I expected a swaggering mountain man with an ego to match his reputation, but when he introduced himself while we were both guiding small groups of brown bear viewers at Pack Creek, I was surprised by his humility, openness, and eagerness to talk about anything related to bears. When I signed on to work for his small guiding operation, I soon learned there was one downside to working for Ken: he called so constantly that MC got jealous.

"Don't answer that!" MC yelled when my phone rang as we lay in bed trying to sleep. Her phone lit up next. "Damn it! I never should have given Ken my number!"

An hour would pass as I sleepily listened to Ken passionately talk about bears and the future. He wanted to know every detail of the bears I'd seen that day, which would lead him off on tangents about behavior and future ideas about the business. With Ken, it was all about what was around the next river bend. He was fearless, or maybe just too excited about life, to be afraid of bears—or seemingly anything, for that matter. It wasn't until the second year I worked with him that I realized how little I knew of his past. Sure, we chatted for what felt like hours most days during bear season, but Ken rarely volunteered any information about his life. I learned to enjoy our long conversations. After all, both MC and Ken's wife, Julie, could only talk so much about bears before wanting to move on to another subject. It was only after the cancer that I began asking for his stories.

I scanned the edge of the muskeg for several minutes, hoping to see the shape of a deer or a flicker of movement. The tracks of a wolf skirted the edge of the meadow. Easing into the old-growth forest, I stared up at giant trees rising into the gray murk and became distracted again with thoughts of my friend.

Ken was born in a Boston suburb and grew up teaching himself how to bird-watch on golf courses. In his seventh-grade journal he wrote, "When I grow up I want to move north and protect wildlife." His life was one of wealth, privilege, and prestige. After graduating from college in 1979, he and a few good friends ventured to Alaska to climb Denali. They spent three weeks skiing in from a train station to the Muldrow Glacier, before spending a month ascending the mountain. When the team was a few hundred feet shy of the summit, Ken's best buddy collapsed over his ice axe with high-altitude pulmonary edema. Ken turned the group around and they began a hundred-mile slog back to the station, running out of food along the way.

After making it back to civilization, Ken traveled to Juneau looking for work. He was hired by Alaska Discovery, a guiding company he'd later own, to move canoes on Admiralty Island. The wilderness of Admiralty soon had a hold on him. He'd spend the next three and half decades guiding there as well as exploring the island on his own.

Ken never had any real problems with bears. He had plenty of close encounters though. One of his favorites involved an Alaska Discovery expedition camped at Swan Point on Admiralty Island. Early one morning, a guide woke to a client calling for him in a low voice: "Are you awake? There's a bear leaning against me."

The guide stood up and was more than a little surprised to see a bear fast asleep and snoring against the client's tent.

"You figure that many trips—around one hundred expeditions annually—every summer over a twenty-year period," Ken reflected. "Bears on every single trip. Never had a bear get food, never had a charge, never had any touching. To think being in wild country with all those wild animals, it shows an incredible aversion of bears to having conflict. People are always asking about bear charging stories. I don't have a lot of bear charging stories."

Ken was one of the first Americans into Russia after the collapse of the Soviet Union. He guided a group of hard-core clients around Siberia and Kamchatka, living with Indigenous Chukchi reindeer herders and traveling by umiak, boats made from walrus hides. He told stories of being abandoned by Native guides with only a walrus head for food, killing and butchering reindeer and drinking their blood, being arrested by the Russian military, seeing more than a dozen dead whales hauled up on a spit to feed foxes bound for the fur market, becoming so sick with giardiasis that he almost died. The best story, though, was when a coup occurred, the country floundered in chaos, and all international visas were canceled. Ken had gotten his clients on a plane back to the States the day before, but he was stuck indefinitely in Russia. He took full advantage of the situation when he was granted permission to stay with a biologist living at Kuril Lake, surrounded by the Kamchatka wilderness. There were two to five million red salmon in Kuril Lake and brown bears everywhere.

"It was just crazy," Ken remembered. "It felt insane to be in that dense of an area." Generally, a female will have two cubs, but Ken recalled, "Three cubs a sow was common. Four, not rare." Such a high number was indicative of the abundant food supply. "Bears everywhere jockeying and growling to get fishing holes. The biologist by himself with no gun or bear spray. They didn't allow guns in the Soviet Union."

The biologist, who had no funding from the government, was the only person trying to prevent poaching in the area. His backpack was torn up from when a bear had "roughed him up." During a day hike, Ken and the Russian saw fifty-six bears and were charged twice, by two different bears. Ken glossed over the details of charges—both times, he said, it was bears guarding fishing holes.

I squinted through the foggy woods and slowly hunted from one bench to the next. A few years ago, I had found the skeleton of a bear near this place. Most Native cultures forbid a hunter to bring the head of a bear home; many will leave it ceremoniously placed somewhere like a nook of a tree. But I'd wrapped the skull in my jacket, brought it home, and gifted it to my dad. Now, I wondered if I should have left it where it lay.

A moment later a deer appeared fifty yards away—a two-year-old doe. When I was a teenager I'd made a weak promise to never shoot another doe, but today I was willing to break it. It wasn't a good time of year to hunt deer, so given the circumstances, I had decided to take the first adult deer I came across. She looked at me, ears flicking. Gently, I worked a bullet into the chamber. I wrapped my rifle sling around my forearm to steady my aim and put the crosshairs on her chest. She watched, unmoving. Then I lowered my rifle and whispered, "Walk away."

She stared at me for a half minute. I raised the gun again, flicked the safety off, and held on her vitals.

"Walk away," I said, this time louder.

I lowered my rifle. She took a step toward me and stood broadside.

"Walk away," I said and then brought my rifle to my shoulder and pulled the trigger.

When I came upon her, she was lying atop moss and ferns, her eyes clenched shut. My hands trembled, but the only solace I could offer was to quicken her death. I took her home, ground her flesh, and brought it to my friend.

At the end of March, Ken invited me over for what he called his "fantasy dinner," which consisted of the two of us sharing slow-cooked caribou. We had one last conversation about bears, guiding, and wilderness. Ken reflected on one client, an older gentleman, who had stood out during his career. They were sitting somewhere in the Arctic, and the light on the mountains was beautiful.

"He expressed it more eloquently than I can, but he basically spoke of his gratitude for God—not just being in a wild place but being created as an aesthetic being to appreciate it. For me, it's always been huge, quiet landscapes. To be able to travel and live a daily existence in wild places. I never really vacationed outside of Alaska. When we did our last trip out to Admiralty last summer and we went back out to the estuary for that last hour waiting for the plane—just leaning against the log, having it be quiet, no clients around to explain to, and there were bears around but I wasn't really focused. It was more just the power of realizing that decade after decade, for hundreds of years going back and, hopefully for hundreds of years going forward, this scene just keeps repeating. This scene of fish and bears and mountains and fog and rain."

On April 11, just as the first bears of the year were emerging from their dens, Ken passed.

A FEW WEEKS LATER I WAS ON KODIAK ISLAND DOING A WALKABOUT WITH my friend Chris Miller. From Larsen Bay, on the northwest side of the island, we hiked and floated down the Karluk River in packrafts to the village of Karluk. We stared out on the calm seas of Shelikof Strait and could faintly make out the coastline of Katmai where Treadwell once enacted his strange drama. The infamously stormy strait was named after

Russian fur merchant Grigory Shelikhov, who arrived at Three Saints Bay on the south side of Kodiak Island in charge of two heavily armed ships in 1784. After two decades of the Indigenous Koniag people repulsing Russian fur hunters, Shelikhov was determined to establish a permanent colony. The Russians slaughtered hundreds of Native men, women, and children. When the cannon smoke cleared, Shelikhov had subjugated an entire people and established the first capital of Russian Alaska. Under Shelikhov's rule, Karluk became a Russian outpost in the late 1780s.

The silent village felt heavy with ghosts. Derelict canneries dotted the bay. Once, more salmon were caught here than in Bristol Bay, but those runs were a ghost of what they had once been. On an eroding bluff, above old houses and shacks gradually falling into the ocean, stood the Ascension of Our Lord Russian Orthodox Chapel. Believed to be the oldest existing church in Alaska, it too will soon be claimed by the eroding bluff and ocean. I walked through the graveyard—it seemed everyone buried here had died young—and was happy to climb over a mountain and stare out on wild country.

Late one evening, we pitched our tent on a bluff above the ocean. Nearby, two does, heavy with fawns they'd soon birth, grazed yellow grass. A mated pair of tundra swans flew in circles before landing on a nearby pond. The sun bit red across the snowy mountains of the Alaska Peninsula and the seemingly endless expanse of ocean. I studied a grandfather trail that had been used for generations, each track a hole several inches deep. It disappeared at the edge of an eroding cliff above the ocean. The trail came from mountains where I knew bears were stirring in their dens and being reborn from the earth. I glanced out at the dimming tundra and watched for movement in the long twilight.

The land came alive with bears in the ensuing days. There were no close encounters or nocturnal visits. There were just bears going about their lives, mostly unaware that Chris and I were watching. Male bears prowled the woods and snowfields searching for females in estrus. Mated pairs rested and foraged on mountain slopes. Subadults dug roots and lazed in the spring sun. The last bears we saw before we left the mountains were a mother with spring cubs. The mother dug roots high on a ridge as her cubs wrestled nearby. I watched the tiny cubs excitedly exploring their big, new, and dangerous world and thought of Ken.

CHAPTER 16

Trust

Nor can we ask bears to repair whatever emptiness we bring with us into the wilderness.

—Sherry Simpson, *Dominion of Bears*

I first visited Pack Creek in 2011 during a brief pause on a kayak trip around Admiralty Island. It was late April, and most of the bears were still asleep in their dens. I'd heard intriguing stories about the place and its bears. The oddest concerned a woman who saw a bear approaching and dropped into a fetal position to play dead. Supposedly, the bear lay down, spooned her, and fell asleep. I explored the woods, studying the gallon jugs, machinery, and dilapidated structures that Stan Price, who'd homesteaded here for nearly forty years, had left behind. In the meadow was a steam donkey—a tool used in logging to haul cut trees down from the woods to the ocean—mounted atop a big log raft. Price had cherry-picked giant, ancient conifers, selling them on the market for good money. He also had worked small gold mines nearby. The weather was calm, and I had a long way to go. I paddled away without the slightest premonition that I'd spend a significant amount of the next decade guiding bear viewing trips here.

Pack Creek was designated Alaska's first bear viewing area in 1934, though it wasn't actively managed until the late 1980s. The watershed and about a mile of beach were closed to bear hunting. Because they had been relentlessly shot during the previous decades, bears had learned to stay out of the open and in the relative safety of the rainforest. The Civilian Conservation Corps built a mile-long trail through the woods, then a viewing platform in a large spruce tree above salmon spawning beds, where people would have a better chance of seeing a bear.

In the early 1950s, Stan Price and his wife, Edna, came to Pack Creek, fleeing the IRS and a civilization that left little room for rugged individuals who wanted to carve a living out of the land and not be beholden to anyone. Legend has it that shortly before the Prices arrived, IRS agents confronted Stan at a mining claim he owned at Windham Bay and threatened to repossess his bulldozer. Stan drove the machine down onto a tidal flat and let the rising ocean ruin the engine. He told the tax collectors it was theirs to take, and then he and Edna left on their boat, towing their float house behind. Most stories of Alaskan homesteaders from the 1950s and earlier involve a zero-tolerance policy toward bears. The Prices, however, tried to get along with the bears of Admiralty Island. It was a novel idea at the time, and one that would eventually elevate Stan to legendary status.

Stan inspired strong feelings in those who knew him—there are differing opinions and contrasting stories concerning his character. What isn't disputed is that by being a regular, somewhat neutral presence at Pack Creek, he helped a number of bears become more tolerant of people. In the process, he taught many people to be more tolerant of bears. He and Edna even raised a few orphaned cubs. People came from around the world to visit the "bear man" and his bears. Visitors reported bears sleeping in Stan's woodshed, a mother nursing her cubs on the front deck of the float house, a bear sticking its head through an open window while Stan was spinning a yarn or eating a meal—the stories went on and on. Many well-known Southeast Alaskan wildlife photographers were introduced to brown bears by Stan at Pack Creek.

In the 1970s, as Pack Creek experienced an influx in visitors who wanted to see bears, a similar trend was occurring several hundred miles west at McNeil River State Game Sanctuary in Cook Inlet. McNeil River's waterfall attracts dozens of brown bears when the salmon are running—during one season 144 individuals were observed. At the time, McNeil had no management, or homesteading bear advocate like Stan, to keep people from scaring away bears or getting in trouble. In 1976, Fish and Game sent Larry Aumiller to manage the sanctuary. Working with area biologist Jim Faro, Aumiller developed a system that would call into question long-held beliefs surrounding our relationship with brown bears. Aumiller, described by most as soft-spoken and humble, realized that many bears can learn to tolerate humans if we behave in a predictable, peaceful, and respectful manner. This means traveling the same trails, using the same areas, yielding to bears, and making sure they don't associate us with food. Aumiller came up with a concept of habituation, which he defined as taking away the fight-or-flight response in a bear—key for developing trust between our two species. Over the next thirty years, Aumiller would guide more than 6,000 people and log more than 60,000 bear encounters. Never was a bear shot nor a person hurt by a bear. Many bear viewing areas across the state—including Pack Creek, when rangers began managing the observatory toward the end of Stan's life—adopted Aumiller's system and experienced the same success.

IN 1984 A BEAR AND HER BROTHER WERE BORN IN A DEN ON A MOUNTAIN near Pack Creek. The cubs were blind, nearly hairless, and each weighed a pound. They clung to their mother's chest and suckled milk, rapidly gaining weight and growing fur. By early May, when the family emerged, the cubs each weighed around ten pounds. The mother ate the roots of skunk cabbage to help get her digestive system working again—she had not defecated or urinated since the previous October, when she'd entered her den. The family wandered down to the ocean, grazed new shoots of sedge grass, and dug clams. Stan, then in his mid eighties, was living alone in his float house at the edge of the tidal flats. At first the cub and her brother were nervous, but their mother taught them the man was harmless. The cubs watched as Stan cut wood, worked in his garden, and tinkered on equipment. Two bear researchers, perched in a tree stand, wrote in their notes, calling the female cub C and her brother B. Other people came to the estuary by plane and boat—sometimes ships full of tourists. The old man would lead visitors down to the creek to watch the bear and her family as they fished for salmon.

During the third May in C's life, a switch went off in her mother. A day prior, the sow would have been willing to fight to the death for her cubs. Now, she ran her young off and threatened them when they tried to return to her. Male bears swaggered in pursuit, jockeying and fighting to court the mother. The cubs, now technically subadults, spent a lot of time near Stan's homestead. Most adult males and more aggressive females, the greatest threats to the little bears, rarely tolerated the old man or other people. By his fourth or fifth year, C's brother, like most males, couldn't stand being near people and wandered away from Pack Creek. C grew into a beautiful bear, with light brown fur and white-tipped ears.

In 1989, when Stan was ninety years old, he spent his last summer on Admiralty Island. He watched bears and rain clouds swirling across the mountains. The tourists left. The rangers who'd recently begun monitoring Pack Creek left. Even C left, climbing into the mountains to find a den on a steep slope beneath a big tree. Stan remained until the days grew short and the calling of southbound geese echoed across the ocean. That winter, as C lay in a deep sleep, Stan passed on in Juneau. In the spring, his friends and family returned his ashes to Pack Creek. During the ceremony a bear approached, lay down on the mudflats, and watched the gathering.

Sometime after 1990, a ranger gave C the nickname Patches. Hundreds of people watched her each summer, delighting when she plucked a salmon from the stream or when she chased a smaller bear. Some may have seen her courted, perhaps even mated, by male bears in the spring. They oohed and aahed over her cubs. She became one of the most seen bears in the world. Images of her appeared in magazines and books. She even became the star of a nature documentary or two. A Pack Creek visitor who worked for Pixar dreamed of making a cartoon about her.

During the first bear viewing trip I guided, I kayaked to the south spit of Pack Creek. Patches was nearby digging clams. Though she was then in her midtwenties and still robust, it quickly became apparent there was something seriously wrong with her. Her femur was broken, and there was a nasty wound where the broken bone poked out of her thigh. On closer examination, I saw that her nose was broken and torn. There was a hole in her muzzle that appeared to have been made by the canine of a big bear. When she drank water, liquid drained out of the wound. She hobbled around, dragging her hind leg behind her, while other bears, mostly subadults and mothers and cubs, steered clear. None of us expected Patches to last long. Infection, starvation, an attack from another bear—the wilderness has a way of eating the weak or unlucky.

Each May I returned to Pack Creek expecting not to see Patches. But before long she would appear, limping along the beach. She wasn't charismatic like many of the other bears. Some viewers cringed as they watched her. Some asked why someone didn't do something to help her. Often she would lie down in a bed of clams, in a sedge patch, or behind a rock in the creek and slowly feed. She angered easily—a small bear that came too close or another bear catching a salmon would often elicit a three-legged charge. The seasons passed and, withered by time, Patches began to shrink. Her movements became slower and even more pained. I watched her and wondered what exactly I, and my clients, wanted from being close to her and her kind.

The bear viewing company I work for is partially owned by Kootznoowoo, the Native corporation of the village of Angoon. Some Kootznoowoo board members expressed their unease with the industry. Being close to bears was seen as disrespectful and went against everything a Tlingit person was taught about proper conduct toward the animal. Dan Kirkwood, who took over management in 2017 after Ken passed on, invited board members to join me on a trip I was guiding. He thought that if our Tlingit owners saw how respectful we were toward the bears, they'd feel better. Thankfully, at the last moment they decided not to come. That day I had two lawyers from San Francisco. We watched a huge male bear eating sedge grass for a while, but the men rapidly became bored. They began to joke and curse at the bear, telling it they were going to "open up a can of whoop ass on it, unless it does something interesting" and that "it's a freeloading son of bitch that should try working for a living." I told them the bear could hear them, but that only encouraged their jeering.

It was times like this that made me question bear viewing. Each season it seemed clients became increasingly obsessed with wanting to be close to a bear, often even expressing their disappointment if they didn't have an encounter close enough to fill the screen of their smartphone. I asked other bear viewing guides what they thought their clients wanted from bears. Many had a hard time answering the question. One said clients are emotionally bored and want to feel something. Another spoke about bears being a dangerous, attractive taboo and of our deep fascination with ancient monsters paired with an experience-driven market.

"Photos are social credibility," he continued. "For those that choose dull lives for money, the money is for proving they're doing great things. We have short, often dull lives—and a need to prove we're making the most of it. Being close to a wild bear is the real deal."

Ken Leghorn had said, without missing a beat, "I don't question what clients' motivations are. I'm just happy they want to see bears."

I SHOWED UP AT PACK CREEK IN THE SPRING OF 2018 AND, AGAIN, WAS surprised to see Patches, now thirty-four years old, walking in our direction. I sat on the beach with a couple of clients as she slowly limped toward us, her ribs showing through her bleached fur. She paused at fifteen yards and, for the first time I remembered in eight years, made eye contact. My people were thrilled, but all I saw in her eyes was pain, hunger, and animosity. A few days later, Kirkwood was sitting at Pack Creek by himself when Patches huffed and bluff charged him. Later that season, she would react aggressively without apparent reason. It seemed her mind was drifting, like she was confusing memory with reality. The salmon run was tiny, and the berry crop failed. When Patches was visible, she spent most of her time resting or slowly limping down the stream, on occasion dragging tapeworms several feet long behind her. She avoided other bears, often becoming frightened even when the bears approaching her were small subadults. Sometimes she would stare off into space with a look that reminded me of my grandmother, who had suffered from dementia.

In August I took two elderly women from the UK out on a trip. At first, seeing how seemingly physically incapacitated they were, I thought about canceling. Reluctantly, I helped them get ready, and we flew out to Pack Creek in a small Cessna floatplane. The women described themselves as "ramblers" and frequently traveled together. Both had outlived their husbands and were resolved to take full advantage of the time they had left. They watched in awe as brown bears peacefully went about their daily lives. At the end of the day, the women stared up at the mountains and out at the ocean. After several moments of silence, one said, "I can't tell you how much of a privilege it is to be here." The other agreed.

Most of the wild places ecologically rich enough to contain large predatory animals are gone or rapidly vanishing. We may think we've forgotten our primordial past, but something deep in us—call it DNA, call it subconsciousness—has not. To see a brown bear—once revered as a god, an enemy, and even a family member gone wild—is more invaluable than most of us can fathom. Even though bears want nothing to do with us, we need them in a way we're only beginning to understand. That need, and its resulting demand, will continue to grow as our lives become increasingly industrialized. One day soon, bear viewing excursions could be more profitable than timber or dirty mines built in salmon watersheds. Why do people want to pay a lot of money and travel from all over the world to see bears? On the surface it's to get a photo, attain some sort of social credibility, fill an emotional or spiritual void. But on a subsurface level,

I wonder if it's an attempt to remember something vital about ourselves we don't even know we've forgotten.

Near the end of the 2018 season, I sat watching Patches as she lay near the creek. She painfully got to her feet and began limping along the bank until she disappeared into the forest. I looked over to my clients, who were scrolling through pictures on the viewfinders of their cameras. I gazed up at the mountains covered in the swirling fog and then back down to where the creek entered the woods.

"That bear has a story to tell," I said.

A woman nodded, then went back to going through her camera's images and smiling to herself.

CHAPTER 17

Brotherhood of the Bear

People always say bears smell bad. She smelled good. She smelled like the woods.

—Forest Wagner, on the brown bear that attacked him

In Le Regourdou in southern France, there's a 70,000-year-old grave that many people believe shows evidence of a religious ritual involving brown bears. The body of a young Neanderthal man had been placed by his people on a bearskin in a stone-lined pit. Around the skeleton are some stone tools and a lot of brown bear bones, as if bear and man had been buried side by side. Nearby stone containers hold bear bones, carefully arranged in neat patterns. Some archaeologists claim these arrangements of bones are a coincidence. They point to the evidence that remains of bears have rarely been found in Neanderthal camps outside of caves where bears would have hibernated.

In the spring of 2016, after Forest Wagner was attacked by a brown bear in the mountains above the Chilkat River, media networks from around the world seized the story. Pictures of Forest were all over the *Washington Post,* as well as Yahoo! News and other online media outlets. Not wanting to have anything to do with sensationalizing the story and the further demonization of bears, Forest refused all requests for an interview. The only details made public—and out of respect to my friend, the only information that I will share—were that Forest had been leading a college mountaineering class and had gone alone ahead of his students to scout a safe route down Mount Emmerich.

He'd found a good descent line and was returning to his students when he inadvertently skied over or near a den inhabited by a mother grizzly and her cub. The snowpack was very thin that year, which likely led to the bear being awake much earlier and significantly more stressed than normal.

In early June of 2016, I flew to Anchorage to visit Forest where he was receiving medical care. Despite his large open wounds and crushed bones, he insisted we climb a mountain with 5,000 feet of elevation gain, much of it without a trail. He carried a portable wound-vac around his neck to suck drainage from his wounds. If he had taken even the most minor fall, there was a good chance he would end up back in the critical care unit in the hospital. I followed as he tottered up steep heather and over boulders and scree until we stood on the knife-edge of the summit. I watched my friend in the swirling clouds as rugged mountains came in and out of view in the gray.

When Forest was finally able to return to Juneau in August, after spending four months recovering in Anchorage, I tried to take him to Pack Creek. I felt like it was important for him to be around bears and that it might help with the healing process. At first he seemed excited, but on the morning we had planned to leave, he admitted that he didn't need to see a brown bear that day. We postponed the trip.

Three years later, Forest and I visited Admiralty Island looking for bears. It was mid-May, and light filtering through tree boughs illuminated the forest and understory in a soft green glow. We hiked a bear trail to a giant spruce tree composed of three twisted trunks, a landmark I visit every year, then followed a grandfather trail through thimbleberry bushes back to the beach, and watched a young blond bear wander our way. When she paused to dig clams, we shouldered our packs and began hiking north. We walked along mudflats, giving the edge of the woods where bears like to bed plenty of space. Both of us carried .44 Magnums—Forest's single-action pistol was so old and big, he joked it had belonged to Wyatt Earp.

We rounded a corner and saw another bear digging clams. She rose on her hind legs and studied us, then ran a short distance to the edge of the woods and sat in a patch of sedge grass. A short while later, we saw two big males feeding on sedges in close proximity to each other. The larger of the two looked more tank than animal. The other bear watched us for a few moments before lying down and nibbling grass. More and more bears appeared. A dark, rangy subadult crossed the tidal flat behind us, while others dug clams and browsed grass near the edge of the forest. A large male emerged from the woods and began swaggering in our direction. At seventy yards he smelled us, turned, and charged back into the brush.

Near the head end of the bay, we sat in the grass and watched bears and deer milling and feeding. A Shiras bear crept along the edge of the forest, slowly grazing patches of sedge grass. I'd lost count at somewhere between twenty-five and thirty—and none had been moms with cubs. In all my life, I'd never seen so many bears in one day. At dusk, we inflated packrafts and floated down a stream, portaging around a bear that sat near the creek watching us, out to the ocean. We paddled over calm seas and through darkness until we found a good beach on a small island.

That night, as we ate dinner, we talked quietly about fear, trauma, and the ways in which we try to come to terms with them. I lay awake for a while listening and thinking back to when Forest and I were kids and I borrowed his truck to go explore the Brooks Range. Near the end of a two-week walkabout, after being repeatedly charged and almost knocked down by a grizzly, my nerves were beyond frazzled. The night after the bear encounter, I was camped in a mountain pass and had, perhaps, the most powerful dream of my life. In it, I was in a dark bedroom, sitting in a swiveling chair. I could see a crack of light seeping around a bathroom door and hear something or someone brushing their teeth. Gradually, I realized the thing in the bathroom was death getting ready for me. I knew I had to get ready for it. A voice that sounded like an evil version of David Attenborough began to narrate what was happening while an unseen audience oohed and aahed. Death turned out the bathroom light and I was suddenly alone beneath the stars.

"What do you have to say for yourself?" the narrator snarled.

Out of the night's black canopy, the open jaws of a bear as vast as the universe appeared. Slowly, the jaws descended as the narrator asked me that same question over and over, a jeer in his voice. Right as the jaws engulfed me, I bolted upright awake in my tent and screamed, "I loved!"

IN THE MORNING, FOREST AND I WATCHED A BIG, DARK FEMALE BEAR while she dug, pried open, and ate clams. Green-winged teal ducks and brant and Canada geese flew by and rested on muddy tidal flats. Massive flocks of surf scoters and goldeneye ducks took flight, bringing the otherwise still ocean to life as we pulled up onto the south spit of Pack Creek. The only bear out was Patches, now thirty-five years old, who lay asleep seventy yards away in the meadow. She had defied my expectations again and made it through the winter. We watched her breathing, her ribs protruding from her bleached, gray fur as her lungs heaved and fell. Forest studied her for a while and then lay down in the grass and closed his eyes. The old bear shifted onto her back, blindly staring up at blue sky for a few moments, and then drifted back to sleep.

I never saw Patches again. As far as I know, the afternoon Forest and I spent with her was the last time anyone saw her. The season progressed, and I kept waiting for her battered form to slowly emerge from the woods or limp down the beach. She had been at Pack Creek clamming and eating sedge grass each spring and fishing for salmon in the late summer every year of her life. After a month passed, then two, then three, then four, I knew she was dead. Other bears I'd watched for years failed to show up— some had died due to the stresses of climate change. Others had likely moved on to make a living in other watersheds.

Pack Creek felt empty without the old bear. She had spanned eras, and with her gone, a part of history was too. I couldn't shake the feeling that it was time for me to leave this place—that I needed to do something else with my life. A smell emanated from one spot in the woods. A ranger wondered if it was Patches and suggested that in September, when the observatory was no longer managed, I go and see. I thought about tracing her skull with my fingers and examining her broken yellow teeth, skeleton, and decomposing fur, but it seemed better that no one ever found her remains.

THAT AUGUST FOREST JOINED ME IN BEING PART OF A FILM PROJECT advocating for the preservation of the last stands of old-growth forest of the Tongass. The Forest Service, directed by President Donald Trump and encouraged by Alaska's governor and congressional delegation, had decided to axe the Roadless Rule, which would open more than nine million acres of the Tongass for roads, development, and clear-cut logging and would destroy critical habitat for brown bears. We'd picked one of the most "beary" mountains on Chichagof Island for the film. We wandered grandfather trails and talked to a camera about the importance of wild places and bears. Some might think it ironic that a bear attack survivor would become an advocate for bears. Forest put it simply, "Even surviving this, I don't want to go kill bears. Beyond being beautiful, they're rightfully there. I see them as an ultimate example of a complete ecosystem."

At the end of the shoot, we boarded a small floatplane and flew across Chatham Strait and over Admiralty Island, toward Juneau. I watched my friend staring out the window and thought back to that last day with Patches. I remembered how on the eve of the old bear's death, she lay in close proximity to Forest, and both seemed at peace with each other and the world.

CHAPTER 18

The Grandfather Trail

If we can't save grizzlies out of altruism, we must save them for the not-so-simple fact that we cannot live without them. Our fates—the clever modern human and our ancient companion in the fur coat—remain as mingled as the bones in the cave.

—Doug and Andrea Peacock, *In the Presence of Grizzlies*

In the spring of 2018, I decided to traverse Chichagof Island. My plan was to begin in Tenakee on the east side of the island and spend a week—or however long it took—and walk to Elfin Cove on the northwest side of the island. The morning I was supposed to fly out, I became so anxious, I couldn't stop vomiting. Nerves are a part of the beginning of every wilderness trip, especially solo ones, but something seemed especially wrong. MC had to pull the car over multiple times on the drive to the airport so I could throw up. I wrote it off as getting older, softer, and more cowardly, but the truth is that of the ABC Islands, Chichagof scared me most. Maybe it was because much of it had been logged, leaving the land scarred and angry. Or because of the island's stories of the Tlingit boogeyman, the Kóoshdaa Káa, a spirit that preys upon people and robs them of their humanness. The island also is loaded with brown bears. The population estimate is similar to Admiralty's—some guess there is a bear for every one of the island's two thousand square miles. I almost didn't get out of the car at the airport. Instead, after dry heaving a few more times in the Alaska Seaplanes bathroom, I boarded a floatplane and watched my home disappear as we cut across the ocean toward Chichagof.

The one person I encountered in Tenakee stopped and warned me about bears. A few hours later, I saw my first—a little subadult that ran as soon as it became aware of me. My anxiety had faded, but still, I couldn't shake the feeling that something was different about this trip. That evening I met a dark bear that postured then lowered its head and stared—a challenging, though not necessarily aggressive, display. A big female or a seven- or eight-year-old male, I guessed. I stood my ground and spoke gently. The bear shifted uneasily, then turned and walked into the woods.

That night I pitched my tent beneath giant trees just above the high tide line. I turned on my headlamp and read Doug Peacock's most recent book, *In the Shadow of the Sabertooth*. It's largely about the first people to come into North America and the megafauna they contended with. I put the book down and thought about fear, how it's imprinted in our DNA, and that it is the strongest root in the tree that our emotions and beliefs branch out from. The forest swayed in the wind, and rain dripped through spruce boughs as I listened for a broken branch or a heavy step.

In the gray morning I continued up the inlet, reaching its end just as sleet began to fall. I watched a male bear courting a female, both post-holing through deep snow. He trailed behind her, pretending he wasn't interested while she pretended she didn't know she was being followed. Watching the bears break through the snow up to their chests made me wish I'd brought snowshoes. My plan had been to hike fifteen or so miles over to Idaho Inlet, but I didn't think I was strong enough to get through the rotten snow. I rerouted, cutting across a short portage to Port Frederick, where I inflated a packraft. Deer grazed the edge of sloughs, a family of otters played, and massive flocks of waterfowl called excitedly as I paddled past.

A storm rolled in from the Gulf of Alaska, and the trees creaked and moaned in the high winds. Rain drummed relentlessly as I lay in my wet sleeping bag and struggled to stay warm. I thought of bears prowling the darkness and remembered how, when I'd first begun traveling and camping alone in grizzly country, one night I woke suddenly to the sound of heavy padded steps approaching. A bear exhaled then nosed the tent a foot above my head. It inhaled deeply, moaned, and made other sounds as if it was deliberating to itself. I listened as it walked away and began feeding on a lush patch of soapberries I had foolishly camped next to.

In the morning I post-holed through snow along logging roads. The snow became deeper, and the storm grew worse, but the thought of the Neka Bay hot springs kept me from turning back. I was soaked and shivering when I raced up a bear trail to the tub, only to find the water cold. Soon after, travel got easier thanks to a large bear that had broken trail for several miles. I camped on the edge of Otter Lake in a strip of old-growth forest surrounded by clear-cuts. The Kóoshdaa Káa is commonly associated with the land otter. It supposedly looks like a cross between an otter and a human, even though it can shape-shift into different forms. When I was a kid I was terrified of the stories. Later, when talking with Tlingit elders, I learned the Kóoshdaa Káa is much more complex than a monster running around the woods that wants to tear you apart or capture you. It's something that's inside each of us—a fear of sorts that makes a person become lost and turns them into something negative they were never supposed to be. Maybe that's the best explanation for what happened to Meriwether Lewis. I stared out through branches at the mostly frozen surface of the lake. A few goldeneyes were doing courtship displays near the edge of the shore where the water was free of ice. The woods were so dark, I could barely make out my tent even though it was just thirty yards away.

In the morning I followed a grandfather trail back to the logging road. The sky began to clear, and I left the snowy road and hiked toward Mud Bay. The watershed was so wild, I half expected to meet a dinosaur. It seemed like the sort of place that would support dozens or more bears, but I saw only two, both eating sedge grass at the edge of the woods. A brown bear hunting guide and his client greeted me at the edge of a big meadow.

"Dr. Livingstone, I presume?" the guide said as we shook hands.

At Gull Cove I visited with outfitter Paul Johnson. One of his clients, an old man from the Lower 48 who had difficulty walking, told me he was worried he wouldn't get a bear. He'd been trying for six days and had only four days left in his hunt.

"The other evening, we saw a big bear, but the guide told me not to shoot. He thought it was a sow. In another day I'll be hard-pressed not to shoot a sow," he said.

I pried boards off a cabin belonging to my friend Sandy Craig. I'd commercial fished with her and her late husband, Joe, for several seasons. Through a lot of fish blood, slime, and beer, they became my Chichagof family. They had retired a few years back, and Joe had passed on unexpectedly shortly after. I sent Sandy a message on my inReach, teasing her that a bear had spent the winter in the cabin but otherwise it looked to be in good shape. I was exhausted but lay awake feeling uneasy for a long time. When I was a teenager, I had spent the night at an abandoned cabin while hiking out from a ten-day solo winter mountaineering trip. There had been a blizzard outside, and I was grateful for the shelter, but I couldn't shake the feeling that there were ghosts trapped inside the walls.

I hiked along the shore of Idaho Inlet, encountering a young bear whose hair stood up on its back when it saw me. When I spoke to it, it slowly walked toward the forest until its anxiety compelled it to charge for cover. I entered the woods and hiked through a forest of cedars and meadows to a low pass that led to Port Althrop. After my trip, I learned that the worried hunter had killed a big male bear a few hours later at nearly the exact spot I'd entered the woods. I had been descending through a forest of giant spruce and hemlock trees, too far away to hear the gunshots.

It was gray and raining hard by the time I emerged into a salt chuck that flowed into Port Althrop. Canada geese, yellowlegs sandpipers, and harbor seals eyed me as I hiked over to the skeleton of a fishing boat. It had belonged to a hermit named Raymond Lee who'd passed on some years before. I never met him, but the Craigs had told stories of Lee—it was rumored he'd had a family in the Philippines slaughtered by Japanese troops during World War II. Lee had sailed all over the world. He survived numerous shipwrecks; once he spent several months floating around in a life raft in the Indian Ocean until a ship rescued him off Madagascar. Eventually he found a home of sorts on Chichagof Island, anchoring his boat in this lonely salt chuck in the summer and spending the winter tied up to the dock in Elfin Cove.

I inflated my packraft and paddled through pouring rain and ocean swells toward Elfin Cove. I surfed a wave onto a kelp-covered beach and then hiked to a muskeg above town. There were numerous fresh skunk cabbage digs and bear tracks indenting the meadow. All the tracks seemed to belong to a medium-sized bear—my guess was a bear that wanted to avoid contact with its own kind. This small peninsula had less-than-ideal feed but was close enough to people to scare away most bears with more aggressive tendencies. My old friends, warmth, and security were just a short hike away, but I wasn't ready to reenter that world yet. I pitched my tent in the gray, watching the ocean roll and the occasional mountain appear in the swirling clouds. I thought about MC and was thankful I had her to return to. I thought about the nearby bear and, despite a whisper of fear, was thankful I wasn't alone in the darkness.

A FEW WEEKS LATER, I SAT WITH MC AND TWO BEAR VIEWING CLIENTS ON a knoll at the edge of the forest at Pack Creek. A mated pair of bears were resting on the beach nearby. That day, we had watched three different sets of bears spread out across a meadow and tidal flats, all copulating at the same time. Mating season peaks in late May and early June. Both males and females tend to breed with a number of bears. A mated pair may stick together for hours, other times for days, and supposedly, occasionally for months. The male, a seven-hundred-pound bear, sat leaning back, watching the female as she napped. He was one of the few adult males that was comfortable with people. Ken Leghorn had named him Smiley because he had upturned lips that made it look like he was grinning. He generally hung around Pack Creek during the spring courtship and then spent the rest of the summer in watersheds seldom visited by people. The female bear was big, dark, and extremely tolerant of people. A Pack Creek ranger had named her Mocha.

There was something in the air that spring. Shortly after I returned from Chichagof, MC and I learned she was pregnant. I was still trying to wrap my head around becoming a dad and found the idea of bringing a child into the world far more daunting than any wilderness trip I'd made.

Mocha rolled over and rose to all fours. Smiley got up and pushed her to mate, but instead, she swatted and bit at him. Then she ran at us, to use us as a buffer—an attempt to scare Smiley away for a while. There was the possibility that Smiley, being both used to people and the most dominant male to frequent the watershed, might try to run us over. Within seconds, Mocha was feet away with Smiley barreling after. He pivoted to try to cut her off and came over the knoll on a trajectory that looked like it would trample MC and me. A moment later, the giant bear towered over us. I spoke gently, and the bear froze and stared down with his small brown eyes.

"I'm sorry. We're here," I said, aiming a can of pepper spray up at his face in case he kept coming.

Near the end of his life, Ken Leghorn, a normally fearless man, had looked at a photo I'd taken of Smiley in the dark rainforest. He'd shuddered and said, "Can you imagine if that thing decided to come for you?"

Smiley stared at me, then through me. He glanced down at MC. If he wanted to, he could have decapitated us both with one swing of his paw. Instead, the giant bear backed off a few steps, walked around us, and followed his mate into the woods.

It was a rough year for bears in Southeast Alaska. The salmon runs were tiny, and the berry crop failed entirely. That fall I saw bears, desperate for calories, behaving in ways I had never observed. They were all over the woods in unusual areas, eating vegetation that had little nutritional value to try to satiate their hunger. Entire meadows looked like they had been rototilled by bears digging for roots. A bear hunting guide told me of a big male his client killed late in the fall that had little fat on it and an empty stomach. On October 1, a young miner was killed and partly eaten on Admiralty Island. I wondered if the bears would have enough stores to make it through winter, let alone grow, birth, and then keep cubs alive.

I watched MC's belly grow and tried to suppress dark thoughts about the state of humanity. The push to open up much of the Tongass's old-growth forest to clear-cut logging was forging ahead. A multitude of giant dirty mines were being built in British Columbia at the headwaters of transboundary salmon watersheds shared with Southeast Alaska. The permitting of the Pebble Mine was looking more likely. There was a push to build a 220-mile road through the southern flanks of the Brooks Range to create a mining boom in western Alaska. Bringing a kid into a world in the midst of the sixth mass extinction seemed questionable at best. Despite all these real and imagined wrongs, the baby inside MC continued to grow. In late January, she went into labor. After two hours of pushing, the doctor guided my hand around the head of our baby and told me to pull. A gray, slimy, seemingly lifeless body emerged. I fumbled with the boy, and then, as I held his limp body in my hands, a switch flipped and he breathed to life, flushing with color. The boy's first cries sounded the same as a bear cub bawling.

Around the time our son was born, Mocha, in the darkness of her den high in the mountains, was giving birth to two blind, nearly hairless cubs. Their mother had carried them in the form of fertilized eggs all summer and fall. It was likely that at least one of them was the progeny of Smiley—often, each egg is fertilized by a different male. In late October, after Mocha had denned, she was healthy enough that her body allowed the eggs to attach to her uterine wall. Snows came, and her body temperature dropped and her heart slowed to eight to ten beats a minute. The weather periodically warmed, and rains threatened to flood her den—a climate change phenomenon that had begun regularly happening during the last five years. Some bears were forced to move and dig new dens, some dying or losing their cubs in the process. Others would wake long before there was enough food to satiate their hunger. Mocha was smart and had picked a place that had good drainage for her winter sleep.

IN THE SPRING OF 2019, AT PACK CREEK, I SAT WATCHING MOCHA AS SHE and her two cubs approached. I was happy to see the cubs as they wrestled and chased ravens, but I knew the odds they were facing. At twenty-four years old, Mocha had had only one cub make it to adulthood. Cub mortality rate during the first year is high mostly due to predation from adult male bears; one study put Admiralty Island's at 60 percent. Some people theorize that female bears who are more habituated to people are worse mothers. Mocha's behavior was different that spring. She kept her cubs close and possessed an anxiety I'd never seen in all the years I'd known her. I thought how against all odds—the lack of food the previous year, another season of poor salmon runs, dangerous bears, people who want to kill her kind, and threatening new weather patterns—she had brought these cubs into the world. And, in a way, they were connected to my son.

Shortly after I first saw Mocha and her cubs, MC and I decided to take our baby to Admiralty on a paddle trip. The first evening we watched a mated pair of bears traveling along the edge of the forest. The male looked like he was wearing a mask, his fur lighter around his eyes, nose, and jaws. He was uneasy when he sensed us. We waited until he was distracted by his mate and left him to peace. That night, I dreamed there was a bear roaring outside the tent. I was powerless to move. My son lay next to me, staring into my eyes. At any moment the bear would come through the thin nylon wall, take him into the darkness, then tear and swallow his fragile form. I woke, felt for my pistol, and then studied my baby fast asleep in the dim light. He breathed deeply, occasionally snoring like his mother, who lay on his other side.

In the morning, we made breakfast beneath towering trees. I drank coffee as MC sang to our son and nursed him. After he accidentally got dirt in his mouth, we joked that his first solid food was soil trodden upon by brown bears. We paddled north, the boy strapped to me in a carrier, past Pack Creek and up Seymour Canal. A small bear walked along the beach in the distance. It was hot, and most were resting or foraging in the shadow of the rainforest. The bear wanted clams but overheated after digging for a short while and retreated into the shade of the forest. In the evening, we made camp on a beach next to a rock that looked like a giant bear. I built a fire, made dinner, and watched MC and our son as alpenglow crept onto the snowy mountains of the mainland.

In the morning we waited for the tide to change so we could paddle—there are some passages that are possible to travel only during certain times. When the current slowed, we launched the boat. Every hour or two, we pulled up onto the beach so MC could nurse the baby. When we reached the giant tidal flat at the head end of Seymour Canal, MC carried the boy as I lined the boat up a creek. The tracks of a small bear from the previous low tide appeared in the mud. We paused when we saw a mated pair of trumpeter swans. One was hurt, with a broken leg and perhaps other injuries. It tried to move away but stumbled and fell. Its mate called anxiously and refused to leave. We had little choice but to crowd the birds until finally the mate flew away. The injured bird summoned all its strength and, after several hundred yards of flapping and running on one leg along the mudflat, lifted into flight a few feet above the earth. Its mate joined it, and they disappeared above the glimmering sea.

We portaged to Oliver Inlet, where we met a floatplane. I held my son as we cut over forest and past mountains. The ocean glowed and shimmered below. He stared out the plane's window, then up at me, unblinking. We had named him Shiras, after the dark grizzlies of Admiralty Island. When he is older, I will tell him the story of his name. How the government and corporations wanted to clear-cut and build roads across Admiralty. How people even wanted to eradicate the Shiras bear and its kin to make the island safer to be exploited. I will tell him all sorts of people from different walks of life worked together and fought to keep this from happening. I will tell him his name is one of the best stories I know.

I closed my eyes for a short while, listening to the hum of the Cessna, and dreamed of taking Shiras into wild country to help him discover bear trails. I imagined walking in front of him, then behind him, until eventually he would travel the woods alone. I dreamed of him learning to talk and listen to bears, of the places he will go, the things he will learn, and the person he will become during his voyage of discovery. When he is old, fate willing, he will come across a grandfather trail high in the mountains and look out on a world still wild with brown bears.

ACKNOWLEDGMENTS

I received a lot of help with this book and owe thanks to many people. First off, thanks to Kate Rogers and Mountaineers Books for taking this project. This is a book I've wanted to write for the last decade, and I appreciate the opportunity to work with the great team at Mountaineers Books. Thanks to Kirsten Colton, independent editor at the Friendly Red Pen, who helped flesh out structure and ideas. Thanks to Laura Shauger and Ali Shaw for their help with the editorial and developmental process.

Thanks to all the hunting and viewing guides and other folks who took the time to talk bears with me. Thanks to bear attack survivors who shared their stories. I'm especially appreciative for Forest Wagner and the two decades of friendship we've shared. Thanks to Ralph Young, Karl Lane, K. J. Metcalf, Joel Bennett, and many more people who fought for Admiralty Island and its bears.

Thanks to my mom and dad for choosing Southeast Alaska and for encouraging me in my writing and wanders even when it stressed them out. I owe a lot to my dad for introducing me to brown bears and wild places, and even more for teaching me respect and empathy for wildlife.

Thanks to my lady, MC—I don't know how I got so lucky as to share a life with you. This book never would have happened without your encouragement, editing, and suggestions.

SELECT
BIBLIOGRAPHY

Ambrose, Stephen E. *Undaunted Courage: Meriwether Lewis, Thomas Jefferson, and the Opening of the American West.* New York: Simon & Schuster, 1997.

Brackenridge, Henry Marie. *Views of Louisiana: Together with a Journal of a Voyage Up the Missouri River, in 1811.* Pittsburgh: Cramer, Spear and Richbaum, 1814.

Brown Bear of Alaska, Hearing Before the Special Committee on Conservation of Wild Life Resources on the Protection and Preservation of the Brown and Grizzly Bears of Alaska, United States Senate (January 18, 1932).

Brown, David E., and John A. Murray, eds. *The Last Grizzly: And Other Southwestern Bear Stories.* Tucson: University of Arizona Press, 2014.

Clark, Frank. "Old Ephraim." In *Chips* annual publication by Utah State Agricultural College. March 3, 1953. Digitized by Utah State University Merrill-Cazier Library. https://digital.lib.usu.edu/digital/collection/Ephraim/id/82/rec/4.

Dobie, J. Frank. *The Ben Lilly Legend: The Greatest Bear Hunter in History!* Austin: University of Texas Press, 1982.

Dufresne, Frank. *No Room for Bears: A Wilderness Writer's Experiences with a Threatened Breed.* Portland, OR: Alaska Northwest Books, 1991. First published 1965 by Holt, Rinehart, and Winston (New York).

Godfrey, Joseph Charles, and Frank Dufresne, eds. *The Great Outdoors: The Where, When, and How of Hunting and Fishing, Including a Dictionary of Sportmen's Terms.* St. Paul, MN: Brown and Bigelow, 1947.

Gubser, Nicholas. *The Nunamiut Eskimos: Hunters of Caribou.* New Haven, CT: Yale University Press, 1965.

Haines, John. *The Stars, the Snow, the Fire: Twenty-Five Years in the Alaska Wilderness*. Minneapolis: Graywolf Press, 2000.

Heacox, Kim. *John Muir and the Ice That Started a Fire: How a Visionary and the Glaciers of Alaska Changed America*. Rev. ed. Lanham, MD: Rowman and Littlefield, 2014.

Herrero, Stephen. *Bear Attacks: Their Causes and Avoidance*. Lanham, MD: Lyons Press, 2002.

Hittell, Theodore H. *The Adventures of James Capen Adams: Mountaineer and Grizzly Bear Hunter of California*. Boston: Crosby, Nichols, Lee and Company, 1860.

Holzworth, John M. *The Wild Grizzlies of Alaska*. New York: G. P. Putnam's Sons, 1930.

Howe, John R. *Bear Man of Admiralty Island: A Biography of Allen E. Hasselborg*. Fairbanks: University of Alaska Press, 1996.

Jans, Nick. *The Grizzly Maze: Timothy Treadwell's Fatal Obsession with Alaskan Bears*. New York: Penguin, 2005.

Kaniut, Larry. *More Alaska Bear Tales*. Portland, OR: Alaska Northwest Books, 1990.

Leopold, Aldo. *A Sand County Almanac*. New York: Oxford University Press, 1949.

Lewis, Meriwether, and William Clark. *The Journals of Lewis and Clark*. New York: Penguin, 1964. First published 1814.

Marshall, Robert. *Alaska Wilderness: Exploring the Central Brooks Range*. 3rd ed. Berkeley: University of California Press, 2005.

——. *Arctic Wilderness*. Berkeley: University of California Press, 1956.

Peacock, Doug. *Grizzly Years: In Search of the American Wilderness*. New York: Holt Paperbacks, 1990.

——. *In the Shadow of the Sabertooth: A Renegade Naturalist Considers Global Warming, the First Americans and the Terrible Beasts of the Pleistocene*. Chico, CA: AK Press, 2013.

Peacock, Doug, and Andrea Peacock. *In the Presence of Grizzlies: The Ancient Bond Between Men and Bears*. Lanham, MD: Lyons Press, 2009.

Rearden, Jim. *Tales of Alaska's Big Bears*. Prescott, AZ: Wolfe Publishing, 1989.

Roosevelt, Theodore. *Hunting the Grisly and Other Sketches*. New York: G. P. Putnam's Sons, 1902.

Shepard, Paul, and Barry Sanders. *The Sacred Paw: The Bear in Nature, Myth, and Literature*. New York: Penguin, 1992.

Simpson, Sherry. *Dominion of Bears: Living with Wildlife in Alaska*. Lawrence: University Press of Kansas, 2013.

Special Committee on the Conservation of Wildlife Resources, S. Res. 246, 1940.

Storer, Tracy I., and Lloyd P. Tevis Jr. *California Grizzly*. Berkeley: University of California Press, 1996.

Treadwell, Timothy, and Jewel Palovak. *Among Grizzlies: Living with Wild Bears in Alaska*. New York: Ballantine, 1999.

Trimble, Marshall. "Mountain Men: The Men That Left an Indelible Mark on American and World History." *True West*, June 10, 2019, https://truewest magazine.com/mountain-men/.

Wolfe, Linnie Marsh. *John of the Mountains: The Unpublished Journals of John Muir*. Madison: University of Wisconsin Press, 1979.

ABOUT THE AUTHOR

A writer and wilderness and film guide, Bjorn Dihle is a contributing editor for *Alaska Magazine* and *Hunt Alaska Magazine* and has been published in *Sierra, High Country News, Outdoor Life,* and other magazines. He is the author of two previous books, *Haunted Inside Passage: Ghosts, Legends, and Mysteries of Southeast Alaska* and *Never Cry Halibut: And Other Alaska Hunting and Fishing Tales.* A lifelong resident of Southeast Alaska, he lives on Douglas Island with his family.

recreation · lifestyle · conservation

MOUNTAINEERS BOOKS is a leading publisher of mountaineering literature and guides—including our flagship title, *Mountaineering: The Freedom of the Hills*—as well as adventure narratives, natural history, and general outdoor recreation. Through our two imprints, Skipstone and Braided River, we also publish titles on sustainability and conservation. We are committed to supporting the environmental and educational goals of our organization by providing expert information on human-powered adventure, sustainable practices at home and on the trail, and preservation of wilderness.

The Mountaineers, founded in 1906, is a 501(c)(3) nonprofit outdoor recreation and conservation organization whose mission is to enrich lives and communities by helping people "explore, conserve, learn about, and enjoy the lands and waters of the Pacific Northwest and beyond." One of the largest such organizations in the United States, it sponsors classes and year-round outdoor activities throughout the Pacific Northwest, including climbing, hiking, backcountry skiing, snowshoeing, camping, kayaking, sailing, and more. The Mountaineers also supports its mission through its publishing division, Mountaineers Books, and promotes environmental education and citizen engagement. For more information, visit The Mountaineers Program Center, 7700 Sand Point Way NE, Seattle, WA 98115-3996; phone 206-521-6001; www.mountaineers.org; or email info@mountaineers.org.

Our publications are made possible through the generosity of donors and through sales of 700 titles on outdoor recreation, sustainable lifestyle, and conservation. To donate, purchase books, or learn more, visit us online:

MOUNTAINEERS BOOKS
1001 SW Klickitat Way, Suite 201 • Seattle, WA 98134
800-553-4453 • mbooks@mountaineersbooks.org • www.mountaineersbooks.org

An independent nonprofit publisher since 1960

OTHER TITLES YOU MIGHT ENJOY FROM MOUNTAINEERS BOOKS

WILD SHOTS
A Photographer's Life in Alaska
Tom Walker
Memoir by renowned wildlife photographer,
author, and naturalist

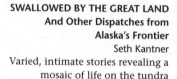

SWALLOWED BY THE GREAT LAND
And Other Dispatches from
Alaska's Frontier
Seth Kantner
Varied, intimate stories revealing a
mosaic of life on the tundra

IT HAPPENED LIKE THIS
A Life in Alaska
Adrienne Lindholm
An open, authentic journey of self-discovery
and learning to find comfort in wilderness

THE STARSHIP AND THE CANOE
Kenneth Brower
Foreword by Neal Stephenson
A timeless tale of a father and son framed by modern
science, adventure, and the natural world

ARCTIC SOLITAIRE
A Boat, a Bay, and the Quest for
the Perfect Bear
Paul Souders
A hilarious, evocative misadventure spanning four
summers and six hundred miles of a vast inland sea

www.mountaineersbooks.org